New Patterns in
Genetics and Development

NUMBER XXI

OF THE

COLUMBIA BIOLOGICAL SERIES

New Patterns in

C. H. WADDINGTON

Buchanan Professor of Animal Genetics

University of Edinburgh

Genetics and Development

COLUMBIA UNIVERSITY PRESS

New York and London 1962

Columbia Biological Series

EDITED AT COLUMBIA UNIVERSITY

Preface

*T*HIS book is based on a set of six Jesup Lectures, given at Columbia University during April and May, 1961. An invitation to contribute to a series as well known as this faces its recipient with a somewhat daunting challenge. He is, I suppose, expected to produce something new; something, moreover, which falls a bit outside the regular well-charted paths of scientific advance of which everyone is already fully aware; and finally, something of his own. My attempt to meet this challenge takes the form of a discussion of two problems, one rather new and one very old. The new problem is the impact of the recent great advances in genetics on our understanding of the development of multicellular organisms. This subject has often been touched on by geneticists but has received less attention from authors who are fully conversant with the embryological material. The old problem is the ancient conundrum of morphogenesis—the appearance of organized structure within a vast range of sizes from the cellular organelle to the elephant. Present-day biology, which is dominated by the enormous successes of biochemistry, has tended to neglect these structures which are too large to be handled by biochemical methods; but they still confront us as one of the most insistent and least understood characteristics of living things. Some may feel that the sensible thing to do at present about these structures is to leave them alone in hope that the progress of biochemistry will eventually throw up some new clue; but to others, including myself, they have the attraction of a real frontier, a region where one is not just trying to fill in an already existing sketch map, but where one has to try to figure out the bare bones of the geography from scratch.

A short set of lectures such as this cannot, of course, attempt to be comprehensive in the treatment even of the topics chosen for discussion, and the desirability of describing some of the work of my own laboratory has led to what would have been an even greater imbalance if I had set out to give a general survey of modern embryology. I have, for instance, consciously left out many important topics on which the main recent contributions have been made by American biologists, since these are matters which scarcely need exposition by a visitor.

There is today in America a great flowering of developmental cell biology as well as of genetics. To an outsider it is perhaps surprising that there is not more contact between the two fields in this country; but I should not like my friends in either camp to feel that the omission from this short series of lectures of many of the topics nearest their hearts is due to any lack of appreciation on my part of the magnificent advances which are being made.

New York, Middletown,   C. H. WADDINGTON
and Edinburgh
March–August, 1961

Acknowledgments

D URING the preparation of these lectures, I had the good fortune to be invited to become a Fellow of the Institue for Advanced Studies at Wesleyan University, Middletown, Connecticut. There I was provided with ideal conditions for thinking and writing, without which it would have been difficult indeed to step back from day-to-day preoccupations and take a look at the problem of development as a whole, as I have tried to do. To the Director of that Center, Dr. Sigmund Neumann, and to the committee who organize the Jesup Lectures, I can only offer my deepest thanks for all the kindnesses and considerations they have shown me.

Authors, editors, and publishers have been generous in giving me permission to reproduce certain drawings, and I should like to express my gratitude to all of them. The name of the author is under each drawing, and the complete reference to his work is given in Works Cited at the end of the book. For the use of the figures, I have been granted permission by the following: *Acta Scientifica Fennica; Australian Journal of Biological Sciences;* Birkhauser Verlag (Basel), *Experientia;* Company of Biologists (Cambridge), *Journal of Experimental Biology;* Elsevier Publishing Co. (Amsterdam), Frey-Wyssling, 1948; *Faraday Society; Genetics; Journal of Biophysical and Biochemical Cytology;* Masson et Cie (Paris), *Archives d'Anatomie Microscopique;* National Academy of Sciences (Washington), *Proceedings of the National Academy of Sciences;* Pergamon Press (Oxford), Waddington, 1959; Ronald Press (New York), Wettstein, 1959; The Royal Society (London), *Proceedings of The Royal Society (London) B.* and *Philosophical Transactions of The Royal Society of London (B.);* Springer Verlag (Heidelberg), *Zeitschrift für Vererbungslehre, Chromosoma,* and *Die Naturwissenschaften,* Kuhn, 1955.

C.H.W.

Contents

Figures

Plates

1. The Production of New Substances

*T*HE title, *New Patterns in Genetics and Development,* which I have chosen for the series of lectures on which this book is based, is ambiguous. It implies, in the first place, that we shall be discussing the problem of form and the appearance of definitely shaped masses of tissue arranged in recognizable patterns, which is one of the most striking and at the same time enigmatic phenomena with which the biologist is confronted. But the arising of orderly forms is only one aspect, and one of the most complex aspects, of the whole process of development. It is usually an accompaniment, and often apparently a consequence of changes in the material constitution of the various regions of a developing system. One can hardly discuss new patterns without devoting a good deal of attention to new substances. And here the other interpretation of our ambiguous title becomes relevant. In the last few years, advances in other fields of general biology, particularly in microbiological genetics and in the ultrastructural biology of adult tissues, have given rise to many new patterns of thought which are applicable, in greater or lesser degree, to the problems of development. The aim of this book is to consider the origin of new patterns of structure in developing cells and tissues in the light of these new patterns of thought.

The changes undergone by developing systems are often spoken of as "differentiation." This is a portmanteau term, and confusion often arises from failure to distinguish the several different types of change which are all included within it; a statement, which may be quite true when one of these is meant, may be nonsense if the word is interpreted in one of its other senses. Differentiation, of course, is always concerned with the differences between two entities, but the entities may be of many different types. The two major categories are differences between two temporal states of the same entity and the differences between two spatially distinct but contemporaneous entities. We find, for instance, that a given region of an embryo changes from gastrula ectoderm through neural plate to neural tube, and finally, perhaps, to the brain. This is a series of alterations in time; and as the name for this general type of phenomenon, I shall use the word "histogenesis." However, we shall find that while one region of the gastrula ectoderm is changing in this way, another region will develop into neurula epidermis and then into the lens of the eye. The difference between the brain and the lens is a difference between two spatially separate parts. For the arising of such differences, I shall use the expression "regionalization."

The differences which arise during histogenesis and regionalization are, to some extent, differences in chemical composition, which could be ascertained in homogenized samples taken from the various stages and regions. But this is by no means the whole of the story. The embryologist is confronted not simply by chemical substances, but by a whole hierarchy of more complex organized entities, such as subcellular organelles, cells, tissues, and organs, in each of which the material substance has some relatively definite spatial arrangement. We need to agree on some terminology in which these spatial factors can be discussed. I propose using the two well-known terms "morphogenesis" and "pattern formation." I shall use morphogenesis when we are concerned with the assumption of a definite shape by a mass of material which we are treating as being homogeneous, that is to say, without separately distinguished parts. When, for instance, the neural plate rolls up into a neural tube, or when a mass of cartilage molds itself into a femur, we can regard these developments as processes of morphogenesis.

They are, in general, only *interesting* examples of morphogenesis in so far as the shapes are definite, that is to say, precisely, or nearly precisely, repeatable in different instances of the same developing system.

I shall use pattern formation for processes in which we wish to distinguish different spatial parts within the developing system and to discuss their geometrical relations. If, for instance, a mass of cartilage in an embryonic limb develops into a number of condensations, and if we wish to consider the relations between the distal and proximal skeletal elements, or the number and arrangement of the digits, then we shall be dealing with examples of pattern formation.

The application of these terms—histogenesis, regionalization, morphogenesis, and pattern formation—overlap to some extent. This is inevitable because the terms are required to discuss various aspects of what is essentially a unified process: that of embryonic development. The best instrument of thought we can hope for, at least at the beginning of such a discussion, is not a set of rigidly defined and exclusive terms according to which we have decided in advance the phenomena should be analyzed, but rather a relatively flexible terminology that uses words each of which emphasizes a particular aspect of the subject without totally excluding the other aspects.

The Nature of Developmental Processes

The first step toward a discussion of these processes should be to decide, if possible, on the nature of the basic concepts in terms of which an adequate framework for our thoughts can be formulated. Chemists conduct their arguments in a vocabulary based on the concepts of atom, electron, and quantum; geneticists find their foundation in the mutation-site and the gene. What are the corresponding fundamental concepts in embryology? A reading of what one might call the classic books on modern theories of development—such as Spemann's *Embryonic Development and Induction* (1938), Weiss's *Principles of Development* (1939), Lehmann's *Einführung in die physiologische Embryologie* (1945)—reveals a discussion couched in terms, such as induction, determination, self-differentiation, competence, regulation, individuation. These are essentially operational terms. They describe, with more or less precision, the behaviors of various types of cells or tissue fragments as they have been revealed by experimentation. It is well known, of course, that there are several difficulties in the exact definition of the various terms; but these are probably no more, or no less, damaging to their usefulness than are the similar complications involved in the meticulous use of other generally accepted and operationally defined physiological terms, such as vitamins, hormones, metabolism. Within their legitimate sphere, the classic embryological terms are capable of doing good service. As all operationally defined terms, they are useful for describing the results of experiments, but are feeble guides, or perhaps even deceptive ones, to the nature of the underlying elements whose properties bring about the processes which the experiments discovered.

What should we take those underlying elements to be? There are, surely, two major clues. In the first place, we know that genes determine the specific nature of many chemical substances, cell types, and organ configurations; and we have every reason to believe that they ultimately control all of them. But, in the second place, the fact that regionalization occurs in the development of nearly all organisms—that in all living things more complex than bacteria (and perhaps even in them also) there are different regions each with its own characteristic specificity—shows that something more than the genes must be involved. This regionalization can usually be traced back to the presence of a number of different types of cytoplasm in the body of the cell from which development starts. Thus, the underlying elements to whose properties we have to look for a penetrating theory of development must be genes interacting with particular types of cytoplasm.

Theories of development based on the interactions of nucleus and cytoplasm go back at least to the days of Boveri. It is only recently, however, that theoretical schemes have been worked out in terms of genes, which we now hold to be the main determinants within the nucleus. In the classic embryological treatises, such as the books mentioned previously, the word "gene" scarcely occurs; and even in such a recent book as the compendious *Analysis of Development* (1955) edited by Willier, Weiss, and Hamburger, the single chapter devoted to genes makes up less than 3 percent of the work. In the 1920s and 1930s, some geneticists —Goldschmidt, Muller, Bridges, Haldane, Garrod, and others—began to be interested in the relation between individual genes and the substances whose specificities they determine, a line of work which has since made such enormous progress. But it was not, I think, until the appearance of my own book *Organisers and Genes* in 1940 that a serious attempt was made to envisage the standard embryological concepts such as competence, induction, determination, in terms of the interactions between groups of genes and their cytoplasmic surroundings. Although among embryologists this interpretation of these concepts is still not generally accepted—at least not in their practice as indicated in books such

as the Willier-Weiss-Hamburger treatise just mentioned—I remain con-
vinced that gene-cytoplasm interactions are the main category of underlying elements to which we can justifiably appeal, although I do not wish
to deny that there may be other types, such as those which give rise to
phenomena of cytoplasmic inheritance which may also play a role
though probably a minor one.

Any differentiated cell contains many different proteins and other
constituent substances. The competence, or determination, of an embryonic cell to differentiate into such an adult type must, therefore, involve the activities of many genes. The groups of gene-action systems
which are concerned with the differentiation of a particular cell type tend
to be interlocked with one another so as to define a stabilized or buffered
pathway of change; these pathways, or "creodes" as I have suggested
calling them, are the basic elements of developmental theory at a level
just more complex than that of single gene-action system—rather as cells
are the elements just more complex than genes in static or nondevelop-
mental theory. But before discussing such complexes, which we will do
in the next chapter, it is necessary to consider how individual genes may
interact with the cytoplasm and with each other. As we shall see, many
new patterns of thought have recently arisen in this context.

We recognize the activity of a gene in a cell by the appearance of
some phenotypic character for which it has been responsible. This phenotypic character may be a specific protein molecule, such as hemoglobin
or an enzyme; or it may sometimes be something else, such as an antho-
cyanin; or a more complex character which is exhibited by a mass of
tissue, such as a bone; or by an organ, such as the tail of a mouse. It is
generally assumed nowadays, and it seems reasonable to do so, that
gene-controlled characters which are recognized as something other than
changes in protein molecules are in fact always the results of alterations
in active protein enzymes. I shall speak of the whole series of biochemical
processes which lead from a gene to the phenotypic character by which
it is recognized as the "gene-action system"; and within each such
system the shorter series of reactions which lead from the gene to the
first protein which it determines will be referred to as the "gene-protein
system."

The regionalization shows that gene-action systems can be and are
controlled. Within an egg which develops in a homogeneous medium,

such as the sea, or inside an effectively closed box, such as an egg membrane, the only candidates for the role of controllers of gene-action systems are the local variations in the egg cytoplasm. We must, therefore, be dealing with systems involving some type of feedback by which the cytoplasm, which itself suffers alteration by the synthetic activities of the genes, also exercises some control over these gene activities.

There is no evidence which forces us to believe that this control ever takes the form of determining the nature of the protein produced by the gene-protein action system. Some of the facts of pleiotropy might seem most easily explicable by such a theory; for instance, such peculiar phenomena as the effect of some *lozenge* alleles in Drosophila not only on the eye facets but also on the tarsal claws (Hadorn, 1955). But in the absence of any biochemical understanding of how such complex effects are brought about, there is certainly no compelling reason to suppose that in these cases the specificity of the gene-protein action system has been altered. We are probably safe in assuming, at least provisionally, that all cytoplasmic control is exerted by altering the rates at which different gene-protein systems operate, and not by changing the nature of the protein which any given gene determines. From the point of view of formulating a biochemical scheme of gene activity, this assumption is a very useful simplification; it will, however, be somewhat questioned in later parts of this book.

The general organization of gene-action systems and their control by the cytoplasm, can be discussed in terms of what I have spoken of as "the double cycle" of intracellular interactions (Waddington, 1954). One of these cycles is a feedback loop by which the cytoplasm affects the genes themselves and controls the extent to which each gene produces its immediate product; the second cycle is a loop by which the cytoplasm affects the gene-action systems at some stage after the production of the immediate gene products. Since we are now dividing the gene-action system into two phases, a gene-protein system succeeded by a protein-phenotype system, the second of the two feedback loops can itself be regarded as composed of two subloops, one affecting the gene-protein phase and one the protein-phenotype phase of the gene-action system (Figure 1).

Before considering in more detail the relative importance of these various possible feedback loops and the mechanisms by which they may

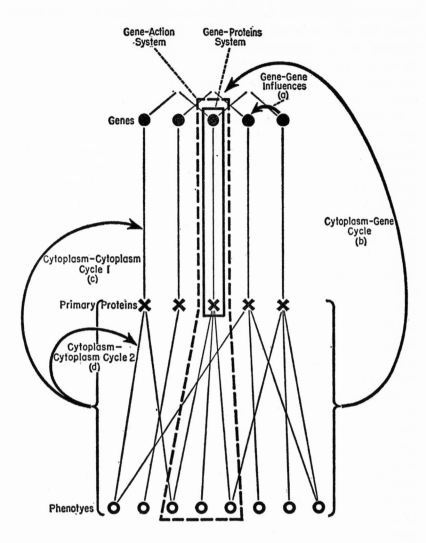

Figure 1. Epigenetic action system of cell

Each gene determines the character of a primary protein, probably through the intermediate steps of RNA and microsomal particles; the primary proteins interact with one another to produce the final phenotype. The whole set of processes connecting a gene with phenotypic character is the gene-action system; the smaller set connecting it with its primary protein is the gene-protein system. Feedback reactions may be from (a) gene to gene, (b) cytoplasm to gene, (c) cytoplasm to gene-protein system, (d) cytoplasm to the primary-protein-to-phenotype processes.

be operated, it will be necessary to summarize some of the recent ideas about the nature of the gene-action system.

Our understanding of the biochemical processes by which genes bring about the synthesis of specific proteins has been, and still is, expanding very rapidly. In fact, knowledge is growing so fast that any account of the subject which attempts to go into great detail is almost certain to be partly out of date by the time it gets published. However, in this connection the embryologist finds himself in a position which, although in general unfortunate, at least has the merit that it partially excuses him from taking the risk of sticking his neck out too far. He is bound to note with regret that the greater part of the relevant work has been carried out on biological materials which do not directly exhibit the processes of development as they are characteristically seen in metazoan cells. For instance, the main work on the mechanism of protein synthesis in metazoan cells has been performed by such techniques as ultracentrifugation, isotopic labeling, and the like, on protein-secreting adult cells, such as those of the liver and pancreas. These studies have been highly informative, but it is well to remember that the basic fact about genes is that they are responsible for bringing specific protein molecules into being, and this is not necessarily the same thing as being responsible for their continued production. The genes, we may say, are architects; and once a manufacturing plant has been set up and put in operation, it passes out of the hands of the architect into those of the manager. It is not entirely inconceivable that in a fully functioning adult liver cell the genes may have long finished their work, shut up shop, and been handed over to the productive machinery which they were originally responsible for building up.

Again, another large section of the most fruitful recent work has dealt with microorganisms, and these differ from developing metazoan cells in two ways, either or both of which may be significant. First, it is difficult, though not in all cases impossible, to find in them any close parallel to the phenomenon of regionalization. And second, during the process of reproduction, the productive machinery of the cells is either handed on more or less intact to the daughter cells, as is presumably the case in vegetative divisions, or at any rate does not obviously suffer the thorough dismantling involved in the formation of gametes from which higher organisms have to make a new start in every generation.

The embryologist concerned with metazoan development must, therefore, exercise some caution in applying to his rather different material the ideas and concepts which his colleagues have developed from their studies on microorganisms and adult cells. At the same time, the students of these other systems have in many cases been able to reach such a pitch of refinement in applying their concepts (often good "old-fashioned" embryological ideas, such as induction by de-inhibition) that their studies are models of the degree of precision which embryologists should strive to equal. What we need to do is to consider some examples of these recent developments, not in order to apply them straightway and without modification to embryological problems, although this may sometimes be possible; but chiefly as examples of the way in which we should attempt to develop into more precise detail the rather general and abstract theories of development which are all that we yet possess.

The Activities of Genes

Current ideas about the gene-protein action system come mainly from biochemical and labeling experiments with adult cell homogenates and the fractions into which they can be separated by centrifugation, supplemented by the genetic analysis of stretches of chromosome in those organisms in which exceedingly rare recombinations can be detected. Perhaps the most unequivocal fact which has emerged from biochemical analysis is that certain gene mutations have resulted in an alteration of only one amino acid in the linear sequence making up a protein which, in the whole molecule, may contain a hundred or more such links. There is almost, but perhaps not quite, as good evidence that the gene itself, or its controlling part, is essentially a deoxyribose nucleic acid which is built up of an ordered linear sequence of nucleotides containing purine and pyramidine bases. Until a few years ago, the analysis by genetic methods of the hereditary materials reached its limit of detail at elements—the classic genes—which would have to contain very many of the nucleotides which are the units in the DNA chain. Essentially all genetic analysis is carried out by one method: the observation of recombinations between elements in two different chromosomes. And the fineness to which the analysis can be carried depends on the rarity of the recombinational events which can be detected in practice. Geneticists who have worked with higher organisms and who have had the patience to observe many more individuals than had previously been considered reasonable have been able to push their analysis below the classic gene and to detect subgenic units in such organisms as Drosophila, Neurospora, and Aspergillus (a recent summary is Pontecorvo, 1959). But these units, nowadays usually referred to by Benzer's term "cistrons," are still so large that they must contain a hundred or more bases of the DNA. It is only in those biological systems in which astronomical numbers of individuals can be handled—that is, in viruses and bacteria—that the genetic analysis can be refined down to the level in which it is detecting changes in the genetic material which one may reasonably suppose to affect only a single nucleotide. However, in such systems, changes of this order of magnitude have been detected, particularly by Benzer (1959).

It seems probable that the hereditary materials of bacteria and viruses differ in some ways from those of higher organisms. For instance, it is not clear if the viral or bacterial chromosome contains protein and DNA as those of higher organisms do; and it certainly seems true that the hereditary determinants in the former can come free of protein in a way that they rarely, if ever, do in more highly evolved organisms. What importance should be attributed to these differences is not clear. Some geneticists who work with higher organisms have argued that in them recombination does not occur within cistrons and that genetic analysis cannot, in principle, be pushed down to the level of the single nucleotide. However, the general view, and the simplest, seems to be that both proteins and DNA are molecules constructed as linear sequences of units (of amino acids in the proteins and nucleotides in the nucleic acids); and, further, that in both cases an alteration of a single link in the sequence produces a change in properties of the molecule as a whole.

Since it is believed that the genes bring about the formation of specific proteins, there must be some sort of relation between the sequence of bases in the DNA and of amino acids in the proteins. But there is no *a priori* logical reason why the relationship should be at all a direct one. The use of one-dimensional sequences is perhaps the simplest way to organize a system which will make it possible to build out of a restricted number of units a very large number of alternative specific complexes. For example, most written language is organized as a sequence of letters forming a word in such a way that the alteration of one letter in the sequence changes the word. A sequence of letters in one language corresponds to a sequence in another language, but the relationship between the two sequences is rarely direct. We can write the sequence G R E A T in English and translate it into German as G R O S S. But if we change the G to a T in the English word, we get T R E A T which cannot be translated into German by a change of the first letter of the German word. There are only a few cases in which there is letter-to-letter relationship between two languages, as when we change English G R E A T into G R E E T and can change the German G R O S S into G R U S S. Therefore, even though proteins and nucleic acids are sequences in which each unit is important, a change in a unit in one sequence does not necessarily produce a correspondingly simple change in the other. It might be that an alteration of one nucleotide in the DNA causes the production of a protein in which the whole amino acid sequence is reorganized.

It is one of the great simplifying hypothesis of present-day molecular biology to suppose that this is not so, but that the sequences of amino acids corresponds in some rather direct and simple way with the sequence of nucleotides. Unfortunately it has not yet proved possible to analyze the sequence of DNA, either by genetic or by chemical methods, in any system in which one can at the same time ascertain the sequence of amino acids in the corresponding protein. Until this can be done, the suggestion that there is a simple relation between the two sequences remains only a hypothesis, although a most attractive one.

The DNA and protein sequences form the two end terms of the gene-protein action system. It is very generally agreed that, even if the two sequences are simply related in formal structure, there is not a direct physiological connection between them, but that intermediate substances intervene. The most important intermediates which have been identified biochemically are ribonucleic acids or ribonucleoproteins. RNA differs from DNA not only in the sugar moiety of the molecule, but also in the bases. DNA contains a sequence of hydrogen-bonded pairs of bases, the pairs always being made up of adenine coupled to thymine, or guanine coupled to cytosine. In RNA, adenine, guanine, and cytosine occur with uracil taking the place of thymine. The arrangement of the bases in the RNA molecule is not as well understood as their organization in DNA, but there are considerable grounds for believing that in some types of RNA the bases form a sequence of pairs similar to that of DNA, while other RNA's have a single-stranded structure built on a simple sequence of bases which are not paired.

RNA fulfills at least two functions and possibly more during the operation of the gene-protein action systems. The first step in the synthesis of protein is thought to be the activation of isolated amino acid molecules to form adenylate complexes with the so-called "pH 5 enzymes." These complexes then become linked to certain RNA molecules, referred to as "soluble" or sometimes as "transfer" RNA. It appears to consist of a group of molecular species; probably there is one species for each of the twenty amino acids which become built into protein. All the sRNA molecules are of relatively low molecular weight (c. 2.5×10^4) and are probably duplex strands. The next step in the synthesis of protein involves another type of RNA known as "template" or "microsomal" RNA. As these names imply, RNA of this type is characteristically found in the microsomal particles which lie in the cytoplasm outside the nucleus. It functions as the templates along which the amino acids (which are in the form of complexes with sRNA) become arranged in a specified order and become coupled together by peptide bonds to form protein molecules. The tRNA molecules are, as might be expected, of high molecular weight (c. 1.7×10^5), corresponding roughly to the size of the protein molecules for which they form the patterns. It has been argued, as we shall see later, that they are single-stranded.

Since it is believed that the DNA genes are responsible for specifying the structure of the proteins and do this through the agency of tRNA, it follows that the structure of the tRNA must be determined by the DNA. However, the tRNA is found mainly in the cytoplasmic microsomal particles while the DNA is in the nuclear chromosomes. There should be some mode of traffic between the two sites. It may be that tRNA is manufactured in the immediate neighborhood of the nuclear genes, and travels from there to the cytoplasmic sites where it carries out the synthesis of proteins. Alternatively, as some authors have suggested (e.g., Volkin and Astrachan, 1957; Jacob and Monod, 1961), the genes may produce a third type of RNA known as "messenger" RNA, which travels from the chromosome to the cytoplasm, carrying the information concerning the order in which the amino acids are to be strung together. The mRNA which these authors claim to have detected has a high molecular weight similar to that of tRNA although more heterogeneous. It differs from the latter in exhibiting a very rapid turnover of bases when offered the possibility of incorporating isotope-labeled molecules; whereas the tRNA in the microsomes has a low turnover rate, and the sRNA also shows little turnover rate except in the adenine and cytidine residues which form the terminal groups in the molecule.

The gene-protein action system undoubtedly involves other substances beside the DNA, the various RNA's just discussed, and the first phenotypic protein which is produced under the guidance of the template RNA. For instance, the synthesis of first phenotypic protein occurs in microparticles in which tRNA is combined with microsomal protein, and these particles are often very intimately associated with other subcellular organelles, such as endoplasmic reticulum or nuclear envelope, which contain lipids and/or polysaccharides. Much less is known about these substances and their functions, but it is well to remember that they exist, for we shall be reminded of them in the next chapter.

With this outline of the elements in a gene-protein action system, we are in a position to consider some of the control systems which are supposed to decide whether or not a given gene-action system will be operative, and at what intensity it will operate. In embryonic systems we know that such control must take place, and we can experimentally switch, in the phenomenon of embryonic induction, a group of cells into one or another of a number of alternative paths of development. Ultimately this switching must involve the exercise of control over which gene-action systems will become functional. It is notorious, however, that the attempts to analyze the mechanisms of these switching processes in embryos have encountered great difficulties.

Nearly a quarter of a century ago, Needham, Brachet, and I (Waddington, *et al.,* 1936) showed that one of the most important inductive switches, that of gastrula ectoderm into either neural tissue or epidermis in amphibia, could be brought about by a substance, methylene blue, which is neither a natural constituent of the embryo nor even something which can plausibly be considered as chemically related to a natural switching substance. We were left with the alternatives of supposing either that the control of the relevant gene-action systems involves only very unspecific substances, which seems almost impossible to believe, or that there are specific controlling substances already present within the cells, and that the externally applied inducers act only to release these substances from some condition of inhibition. This second supposition was our hypothesis of the "masked evocator." We thought, at that time, that we were faced with a technical impasse, and that no progress could be expected from any attempt to determine the specific nature of the natural inducing substance. A similar conclusion was reached some years later by Holtfreter (1945), who in fact argued that there is no specific evocator, the induction-reaction being the result of an unspecific cytolytic action. Actually, as will appear later, Needham, Brachet, and I were being too pessimistic, and Holtfreter was almost certainly wrong; but all three of us who discovered unnatural evocation very quickly dropped the subject and went on to something else: Brachet to his most productive work on the relation between RNA and protein synthesis, Needham to the energy-producing metabolism of various regions of the gastrula, and I to the control of tissue differentiation in Drosophila by systems of genes.

I shall take this opportunity to break off from the direct historical account of the development of the organizer problem in order to introduce one of the newest and most fascinating patterns of thought in this general field. I shall return later to the organizer and show, I hope, that the ideas we will now discuss, novel though they are in some respects, are still an organic outgrowth from the previously existing corpus of epigenetic theory. So let us, for the moment, leave vertebrate embryos for the other end of the biological realm: the bacteria.

The Control of Gene Activity

One of the microbiological phenomenon which seems to offer a close parallel to the fundamental processes of development is the induction or repression of enzyme synthesis. When bacteria are placed in the presence of a substrate suitable for catabolism, it is frequently found that an enzyme capable of breaking down the substrate molecule appears in high concentration in the bacterial cells: this is enzyme induction. Conversely, when bacteria are offered a molecule which they normally synthesize, the enzymes involved in this synthesis frequently disappear more or less completely: this is enzyme repression. This phenomenon is not to be regarded in any way an exceptional one in bacterial physiology; in fact, some authors have suggested that all bacterial enzymes should be considered as potentially involved in processes of induction and repression.

On the other hand, such processes have rarely been found in clear-cut form in metazoan cells, in which only a few instances of them are known; for instance, tryptophane peroxidase can be induced in rat liver cells by the injection of tryptophane into the circulation (Knox, 1954). Attempts to provoke induced enzyme synthesis in embryonic cells have usually been unsuccessful, although Klein (1960) has some evidence for the induction of arginase following culture of chick and mouse cells in arginine-enriched media. It is true that when neural tissue is induced in vertebrate embryos, the cells soon come to contain a characteristically high concentration of choline esterase and probably other unidentified enzymes; but this can scarcely be regarded as a close parallel to induced enzyme synthesis as that phrase is normally understood, since in the embryos the induction has not been carried out by the administration of a substrate for the enzyme which appears. Only in a more general sense, as an example of the control of the synthesis of a specific type of protein, is the induction and repression of enzymes in microorganisms considered to provide a model of the kind of process which may operate in embryos.

Jacob and Monod (1961) have published a rather full discussion of recent work (much of it from their own laboratory) on induction and repression of enzymes in *E. coli,* and have tried to account for the phenomena in an elaborate theory. Without attempting critical assessment of all the details of this, which like all theories is certain to require some modification by other experts in the field, we may take it as a good example of the type of conceptual scheme which is now becoming possible to develop in connection with the control of gene-action systems.

As a typical example of enzyme induction we may consider the formation of β-galactosidase. This is a single protein, usually existing as a hexamer, with a basic unit having a molecular weight of about 135,000. When a suitable substrate is not present in the medium, the wild-type *E. coli* contains 1–10 units of β-galactosidase per milligram dry weight, which amounts to about 0.1 to 1 molecule per cell. In the presence of an inducer, about 10,000 units per milligram dry weight appear. This involves a true protein synthesis rather than a simple conversion of some already elaborated precursor, and is inhibited by substances which suppress protein synthesis. The process starts within a few minutes of the addition of the inducer to the medium and ceases equally rapidly when the inducer is removed. In certain other instances of enzyme induction, for instance, in that of penicillinase in *B. cereus* studied by Pollock (1958), the diluting out of an enzyme following removal of an inducer from the medium is much slower, but this is because some of the inducing penicillin remains in the cells where it continues to act as an inducer.

The reaction of normal *E. coli* to the presence of lactose in the medium involves not only the great increase in β-galactosidase just mentioned, but also the appearance, or increase in concentration, of a permease which makes it possible for the lactose to enter the cells through the cell membrane. This permease has not been as well characterized as the catabolic enzyme; in fact, it may consist of a group of substances, one of which is an acetylase. When induction occurs in normal cells, both permease and β-galactosidase are produced in a constant ratio to each other.

It is clear that induction is a rather highly specific process, since it is only brought about by a very restricted range of substances. However, there are some inducers of β-galactosidase which do not act as substrates for the enzyme, so it appears that the induction does not depend on an exact correlation of molecular structure between inducer and the active site on the enzyme.

A number of mutant types of *E. coli* have been found in which the induction process is defective. In the first place, there are two genetic loci, z and y, mutations which lead in the case of z to loss of capacity to synthesize galactosidase (whether as a result of induction or in any other way) and in the case of y to a similar incapacity to produce permease. These two genes are very closely linked, but are definitely different cistrons, since there is complete complementation in diploids of the type z^+y^- / z^-y^+ which can form both enzymes. One may consider that these are the two "structural" genes, which specify the way in which amino acids are to be strung together to make the two enzyme molecules.

The next element in the system which we need to consider is a locus i. Wild-type cells which contain the i^+ allele exhibit the ability to be induced. In mutant strains with an i^- allele, both permease and β-galactosidase are found as normal "constitutive" enzymes, always present whether there is an inducing substance in the medium or not. The i locus is closely linked to z and y, but is in a different cistron. If, in a diploid, the dominant i^+ allele is present in one chromosome, it can control the activity of z or y alleles in the other chromosome; thus, diploids i^+z^- / i^-z^+ are inducible owing to the combined action of the i^+ and z^+ alleles. This indicates that the i locus affects the z and y loci by means of a substance which escapes from the chromosome (Jacob and Monod speak of it as a "cytoplasmic substance," but it need not necessarily come outside the nucleus). It is most plausible to suppose that this substance is produced by the dominant wild-type allele i^+ and is absent or ineffective in the mutant strains carrying i^- alleles. It is, of course, in the wild-type strains that the z and y loci do not normally produce their enzymes but do so only if inducers are present. Thus, the i^+ substance must be a *repressor* of the activities of z and y; and the action of inducers in the medium must be to annul the repressive action of the i^+ substance; induction is really to be considered as de-inhibition. The i locus, which produces the repressor substance, is spoken of as the "Regulator" of the system.

The mechanism of enzyme repression by substrate seems to be essentially similar to that of enzyme induction. An example is provided by tryptophane synthetase in *E. coli*. Normal cells contain a series of enzymes which control a sequence of steps by which tryptophane is synthesized. If tryptophane is present in the medium, not only the enzyme for the terminal step, but the whole series is suppressed. Indeed, if we have a series of enzymes controlling the sequence of processes $A \rightarrow B \rightarrow C \rightarrow D \rightarrow E$, the presence of E in the medium will suppress the enzyme concerned with converting A to B even in mutants in which the enzymes for the steps $B \rightarrow C \rightarrow D$ are lacking. This shows that the presence of E itself is sufficient to suppress the whole series, and that it need not be converted back to D, and then to C, and so on, to bring about the repression of the enzymes for these earlier steps. Just as in induction a single substance causes the production both of permease and of catabolic enzyme, here a single substance represses a whole battery of anabolic enzymes.

The structural genes concerned with specifying the amino acid sequences in the enzyme of the tryptophane synthetic pathway are, in *E. coli,* grouped together in a linked assembly much as the z and y are. However, this is not always the case. For instance, the structural genes for the enzymes concerned with arginine synthesis are scattered throughout the genome, and not closely linked.

In enzyme systems subject to repression, regulator genes, corresponding to the i locus, can also be found. In the tryptophane case the R (regulator) locus is not closely linked to the group of structural genes; and in the arginine case, in which the structural genes are not themselves linked, the R locus could obviously not lie near more than one of them, and in fact does not lie very near any of the group. Thus, the fact that in the galactosidase system the i locus is close to z and y is to be regarded as contingent and not as an example of a necessary relation.

In repression-susceptible enzyme systems, the function of the regulator locus is to produce a substance which is normally without effect but which becomes a repressor of the activity of the structural genes when it is combined with the repression-inducing substance in the medium. Thus, these external repressing substances are really "co-repressors," which operate after combination with an internal complementary factor.

One must next inquire how it comes about that a presumably single repressor substance, produced by a single regulator locus, can affect the activities of a whole set of different enzymes, such as the permease and galactosidase. Jacob and Monod argued that there must be some substance with which the repressor substance combines in a stereospecific way. To this hypothetical substance they gave the name "the operator." They continued the argument with the consideration that the operator, being a specific substance, must have its constitution controlled by a gene. Now, a mutation of this gene, leading to the production of an altered operator which does not react with and therefore is not affected by, the repressor substance, would put the repressor substance out of action, and thus lead to the appearance of cells in which the permease and galactosidase were constitutive and not inducible; moreover, this lack of inducibility would be dominant in diploids, since so long as a repressor-insensitive operator is present, the z and y genes will be able to produce their products.

A search was, therefore, made for dominant constitutive mutants, and these were duly found. They are referred to as o^c mutants. This confirmation of hypothesis is very satisfactory, but it might appear that we are no further forward in dealing with our original difficulty. This was that the i (regulator) locus produced a nonchromosomal (or "cytoplasmic") substance which had a pleiotropic specificity for both the z and y loci. Again, we find ourselves with a locus, the operator or o

locus, which has pleiotropic specificity for these two structural loci. However, the situation is different in that we have as yet no evidence that the *o* locus acts through the production of a nonchromosomal substance. In fact, the evidence indicates that it does not. In diploids of the constitution o^+z^+ / o^oz^- or o^+z^- / o^oz^+, the *z* allele attached to the repressor-sensitive o^+ gene (z^+ in the first case and z^- in the second) remains repressed, while the *z* allele connected with the repressor-insensitive o^o gene is unrepressed. This shows that the *o* locus operates by some mechanism which is transmitted along the chromosome; and although the precise nature of this mechanism is not known, it is comparatively easy to picture it as affecting a particular group of genes, such as *z* and *y*, not because it has the pleiotropic specificity which would be necessary if it acted on each of the genes individually, but just because they lie next to it on the chromosome. Such a closely linked group of loci controlled by a single operator is spoken of as an "operon."

The complete scheme suggested by Jacob and Monod to account for the control of the gene-protein action systems by which these bacterial enzymes are produced is shown in Figure 2. Admittedly, this scheme is still to some extent speculative and hypothetical. Even the outside observer of this field can see certain points on which he would like further clarification. For instance, it is not clear from the mechanism proposed, why catabolic enzymes should be inducible and anabolic suppressible. Is this merely something which has been brought about by natural selection because such an arrangement happens to be useful? Or is there some more necessary connection between the nature of repressor substances and substrates which makes it likely that catabolic substrates will inactivate repressors while anabolic substrates will act as "co-repressors"? Again, the mechanism suggested for the operator control of operons seems plausible and convenient when all the loci involved in a given inducible or repressible system are closely linked, but somewhat less attractive when they are, as in the arginine synthetase case, scattered widely throughout the genome. We then, apparently, would have to suppose that the operator locus is duplicated, or that there is a separate operator locus for each gene or group of genes. Finally, one has the feeling that the definite, even though not complete, degree of complementariness between the inducing substances and the enzymes produced has got somewhat lost in the complexities of the system. We now have the inducer chemically related to the repressor substance; but the latter is related, not to any of the induced enzymes, but only to the operator. According to this system, the fact that what lactose induces is β-galactosidase, is not an inevitable consequence of the nature of the mechanism of induction, but is presumably the result of the machinations of natural selection, which has made use of a mechanism which could equally well have been employed to enable this sugar to induce any other enzyme for which a structural gene exists.

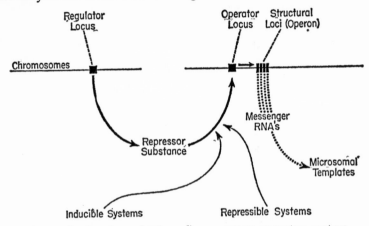

Figure 2. Jacob and Monod's repressor-operator system

In inducible systems the substrates prevent repressor substance combining with and inactivating operator. In repressible systems the repressior substance does not inhibit operator until combined with external repression-inducing substrate.

However, from the point of view of our understanding of development, the details of this scheme are less important than its general character. The most important new patterns of thought to which this example introduces us are, perhaps, two. First, that the genome may contain, in addition to the structural genes which specify sequences of amino acids, at least one if not two types of controlling genes, exemplified by the Regulators and Operators, which can be brought into action in a regular way by external conditions. And second, that the control of a gene-protein action system is frequently—in this example, always—carried out by an influence which affects the gene itself rather than the rest of the system. This second point is a consequence of the existence of operator genes controlling the protein-determining structural genes. The Operator-Operon system is perhaps the most novel component in the whole train of thought suggested by Jacob and Monod, and we shall discuss it first.

We have in the last few years learned of a number of controlling genes, or hereditary elements, in other systems, including higher organisms. Perhaps the best known are those in maize described by workers such as Rhoades, McClintock, and Brink. In some strains, maize kernels exhibit many flecks and spots of color which occur in different sizes and frequencies in different strains. Genetic analysis has shown that these are due to the action of various controlling elements which affect the (presumably structural) color genes. In the strains studied by McClintock (1956a, b), for instance, there are two such controlling elements. No spotting occurs in the absence of one of these, the Activator *Ac*. In its presence a second element, the Disassociator *Ds*, cause the occurrence of chromosome breaks at its position, leading to the formation of dicentric and acentric fragments which become eliminated at some later mitosis in a manner dependent on the dose of *Ac*. The *Ds* element also causes changes in the activity of nearby genes (possibly by bringing about mutations in them) without involving any actual loss of chromosomal material.

Another controlling system, Dotted *Dt*, studied particularly by Rhoades (1941), acts on the color gene *a*, which is quite stable in the absence of *Dt* but in its presence gives rise to numerous small spots of color, presumably due to mutations brought about by the *Dt* element. The Modulator factor described by Brink (1958) is another controlling element, possibly a recurrence of *Ac*. All these controlling elements appear to be genelike in that they can mutate to a series of allelic forms which are recognizable by the particular type of effect they produce on nearby genes. Most, if not all of them, seem to be peculiarly foot-loose entities which are frequently moved from one region of the genome to another. This is sometimes, but not always, a consequence of the induction of chromosome breaks; in other instances, McClintock speaks of the movement as being brought about by some mechanism comparable with the transduction of genes in bacteria.

Students of these controlling elements in maize have frequently suggested that similar factors may control the process of regionalization in normal metazoan development by switching on or off the appropriate gene-action systems (e.g., Brink, 1958). However, it is difficult to see that the systems as they have been described in maize would be adequate to perform this function without rather radical modification. The characteristic of all of them is that their operation is very little controlled. They produce more or less randomly distributed spots and flecks; and although there may be some slight indication of control in that some alleles tend to produce larger and others smaller spots, or spots with greater frequency in certain regions than in others, the systems appear to be very far from providing a mechanism which could bring about the sudden (in terms of cell lifetimes) direction of a whole coherent mass of tissue into one definite line of development, as we see, for instance, when a large region of the gastrula becomes determined to form neural plate. Without discussing, at the moment, exactly what this determination implies, it seems clear that something has occurred consistently in every cell, or nearly every cell, in the whole mass; and this is not at all like the capricious action of the maize controlling elements.

However, even if it seems unlikely that elements of just this kind are active during embryonic development, the maize results are useful in reminding us of the possibility of the control of gene-action systems by other elements in the genome. Moreover, there are some examples in plants in which controlling elements have been postulated of a kind which might perhaps be more easily incorporated into embryological theory. For instance, Lewis (1960) has studied the S locus which controls the growth of pollen tubes in styles of Oenothera. He has come to the conclusion that the locus consists of two cistrons, one of which is a structural gene responsible for the specificity of a pollen tube or style substance, and the other is a controlling locus. With the normal form of the controlling cistron, the substance specified by the structural cistron is produced both in pollen tubes and in styles; but with certain mutations of the control cistron, this same substance is formed in pollen tubes only, with certain other mutations, in styles only. Here we have the invocation of a control system which is sensitive to the differences between tissues; this gets us much closer to something which embryologists might wish to postulate.

As another example of a phenomenon which suggests the possibility of gene-to-gene control—and one which comes still more closely home to students of metazoan development—it is worth drawing attention to the extremely interesting observations of Mechelke (1961). In the second chromosome of the Chironomid *Acricotopus lucidus,* there is a certain region, comprising the bands mapped as P50 to P78, which shows signs of activity in the anterior compartment of the salivary gland during the end of the larval period. This activity seems to be rather directly induced by the ecdysone hormone which is secreted just before pupation occurs (Clever and Karlson, 1960). The signs of activity of these salivary gland chromosomes are the swelling of bands into "Balbiani rings" and the appearance of droplets presumably indicating synthesized materials. The interesting point in the present connection is that these signs of activity appear first in the distal part of the region (bands 78, 77, etc.) at the end of the larval period, and gradually spread along the chromosome to more proximal regions (eventually to band 50) during the course of the prepupal period. Here we have a visible indication of a sequential bringing into operation of gene-action systems lying in series along the chromosomes. This is just the kind of intrachromosomal activity which Jacob and Monod's Operator is supposed to carry out.

We have, therefore, several groups of phenomena which force us to take into consideration the possibility of the control of gene-action systems by effects exerted on the genes themselves. It still, however, seems too early to decide whether all systems controlling gene-action systems have as their last link an influence which impinges on the gene itself, a "genotropic" link as one might call it, using the word construction employed by students of hormones to indicate the site of action of the substances they are interested in. It might well be that in other cases gene-action systems are controlled by, for instance, destroying messenger-RNA's, which are continually being made by the structural genes, or by pre-empting all the available sites on microsomal particles so as to exclude the RNA derived from certain genes; and other mechanisms of control could probably be thought of.

Before leaving this subject, it is worth drawing attention to the possibility that in organisms which are evolutionarily more advanced than bacteria numbers of "Jacob-Monod systems" might have become interlinked (Figure 3). If a structural gene controlled by an operator in the first such system produced a substance which functioned as a repressor substance for an operator in a second system, we would have the possibility of "cascade repression"; and if there were a number of links of this kind, complex systems might be built up which exhibit some of the tendency towards irreversibility which is commonly found in embryological materials but which is hardly accounted for on the simple Jacob-Monod scheme. Other more complicated linkages between different repressor-operator systems are, of course, also possible.

We may now turn to consider the earlier link in the control system suggested by Jacob and Monod; namely, the Regulator genes and the

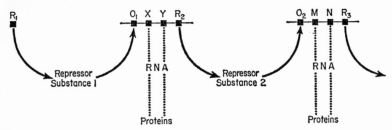

Figure 3. Cascade repression

One of the structural genes controlled by the first Operator produces a repressor substance acting on a second Operator, etc.

cytoplasmic (or at least nonchromosomal) substances which they produce. The idea that there are certain genes which produce substances which regulate the activities of other genes is, of course, a standard part of any embryological theories which consider genes at all. A quarter of a century ago, T. H. Morgan (1934) realized that "The initial differences in the protoplasmic regions [of the egg] may be supposed to affect the activity of the genes"; and also that "all the genetic evidence points plainly to the conclusion that the characters that develop in the protoplasms are ultimately traceable to the genes in the chromosomes [of the maternal tissues]."

Thus the form in which regulator genes have appeared to embryologists is as the genes which control the formation of the differentiated and localized oöplasms of eggs. I have often urged the importance of studying the actions of such genes (Waddington, 1940a); and in my laboratory we have, in fact, over the years, carried out a number of studies on various female-steriles in Drosophila with such considerations in mind (Counce, 1961; Waddington, 1961b). Unfortunately, so far, only a few others have taken up the problem (particularly King and his collaborators, 1957a, b; 1960). None of this work, including even the electron-microscopical studies done to date (Okada and Waddington, 1959; Waddington and Okada, 1960; and see Plates VIII to XI), has proved very informative as to the nature of the substances produced by such genes or their exact mode of action. The crucial question, it seems to me, is the biochemical nature of the cytoplasmic regulatory substances. It is not clear that we know of any very suitable cases among metazoan eggs for attempting to determine this. It might seem that favorable systems would be provided by those female-sterile mutants which cause the production of abnormal eggs which can be restored to normality if fertilized by a sperm bearing the wild allele of the gene. For instance, in eggs laid by homozygous *deep-orange* females in *D. melanogaster,* the union of the pronuclei and the first cleavage divisions are upset unless the fertilizing sperm carries the normal allele of this locus (Counce, 1956). But the type of abnormality in this and in most other of the known cases is not very clearcut and may well be quite secondary to the primary gene action. What the restoring sperm brings into the egg may be not something which acts as a regulator substance, but perhaps a gene capable of producing an enzyme necessary for energy production or some other essential factor in the division mechanism. We have not yet found in higher organisms a gene which alters an oöplasm in such a way as to produce a well-defined (as opposed to a messy) change in the activities of the genes controlling differentiation.

The other approach from the embryological side to the nature of regulator substances is of course the study of evocator substances. As was said previously, Needham, Brachet, and I were too pessimistic in the late 1930s in thinking that the existence of unnatural evocators, which act by de-repressing a natural evocator, would make it impossible to find out anything about the substances which are normally operating in embryos. A new clue came in the discovery that substances can be derived from certain tissues which induce specifically distinct regions of the axial system (Chuang, 1939, 1940; Toivonen, 1940). This lead had been energetically followed up, mainly by Toivonen's group in Finland and Yamada's in Japan, who have been joined more recently by Nieuwkoop's in Holland and the Tiedemann's in Germany.

I do not intend to summarize all this work, which has recently been reviewed by Yamada (1961; see also Waddington, 1959). The work has proved extremely suggestive, if not perhaps fully conclusive, about the chemical nature of the natural evocators. When evocation is brought about by relatively simple chemical substances (abnormal pH or salt content of the medium, methylene blue, steroids, polycyclic hydrocarbons, etc.), it results in the production of neural tissue which either remains as a formless mass or organizes itself into the most anterior parts of the nervous system (the archencephalic part, consisting of forebrain to which may be attached eyes, nasal organ, etc.). Various materials extracted from adult or embryonic tissues may, on the other hand, evocate other well-defined parts of the axis, for instance, the deuterencephalic region (mid- and hindbrain, ear vesicle), the spino-caudal, the trunk-mesodermal, or the endodermal (pharynx, stomach, intestine). These substances extracted from tissues are usually known as "heterogeneous inductors," in contrast to the chemical "unnatural inducers"; and they seem rather likely to be closely related to, if not identical with, the substances active in normal embryological development. (For some stimulating new patterns of thought about natural evocators, see Ebert and Wilt, 1960.)

There are two main questions we have to ask about in the context of gene-control systems. Not only do we wish to know their chemical nature, but also whether they are genotropic, impinging directly on genes as the regulator substances do on the operator genes, or whether they are plasmotropic, impinging on nonchromosomal regulator substances as the inducing substrates do in cases of enzyme induction (Figure 4).

As to their chemical nature, the evidence is that all substances capable of anything more specific than archencephalic evocation are connected with ribonucleoproteins; but of the two parts of such molecules, it seems to be the protein rather than the RNA, which is the active agent (Yamada, 1961). By contrast, the regulator substance (repressor) in the system studied by Jacob and Monod is thought not to be a protein. The evidence for this is that the substance is still produced in the presence of 5-methyl-tryptophane, which normally suppresses protein synthesis; but it is perhaps not quite clear that it would necessarily suppress the formation directly by a gene of some protein of a special kind which might be expected to combine with the operator gene in the manner characteristic of the repressor substance; thus, the nature of the repressor is probably best regarded as still in doubt.

The heterogeneous inducers would, of course, not be expected to be similar to the regulator substances if they should turn out to be plasmotropic rather than genotropic. I have long been interested in the question

of where in the cell the evocators impinge. Some of the unnatural evocators, such as steroids and polycyclic hydrocarbons, fluoresce strongly
when irradiated with ultraviolet, and using this as a means of locating
them, it could be shown that in amphibian embryonic cells they accumu-

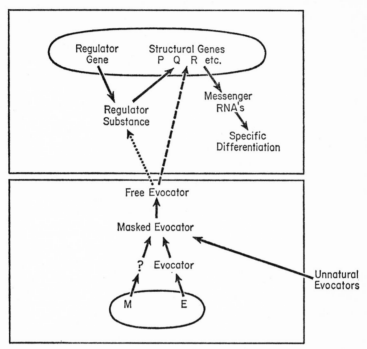

Figure 4. Possible evocator effects on gene-action systems

The lower rectangle represents an inducing cell. In this an evocator substance is
produced, presumably under the control of a gene (E). This may, in some cells,
become masked by combination with another substance, which is possibly also
directly gene-controlled. The unnatural evocators deinhibit the evocator. A cell
competent to react to the evocator is diagrammed in the upper rectangle. The
syntheses constituting specific differentiation are controlled by a Regulator—(Operator)—Structural Gene system. The evocator might act either directly on the
structural genes (or Operator), which would be a genotropic action; or on a
regulator substance, which would be a plasmatropic action.

late in lipoprotein granules in the cytoplasm and do not appear in
recognizable quantity in the nucleus; they, therefore, seem to be plasmotropic (Waddington and Goodhart, 1949).

The evidence about heterogeneous inducers is less clear. The only
way of trying to locate them in the cells on which they act is to label

them with radioactive tracers and use autoradiography. Unfortunately, at the relevant stages of development the most rapid uptake of the small molecular components of proteins and nucleic acids is into the cell nucleus, and since some labeled molecules of this kind must always be present in such experiments, owing to breakdown processes, their uptake into the nucleus will tend to obscure any movement of complex geno-tropic substances into those structures. In fact, what we find is that when unlabeled embryonic tissue is placed in contact with a labeled inductor (either a normal embryonic one or a heterogeneous one) a con-siderable amount of label passes into the nuclei, probably in the form of small molecular breakdown products; but particularly with hetero-geneous inducers, more label is found in the cytoplasm of the reacting cells than is the case when only small labeled molecules are offered (Sirlin, Brahma, and Waddington, 1956; Waddington and Mulherkar, 1957; Sirlin and Brahma, 1959). Thus, the evidence, as far as it goes, suggests that the heterogeneous inducers are plasmotropic; but it is doubtful whether the movement of small quantities of genotropic sub-stances would have been detected. It perhaps is worth pointing out that the substances which act as evocators in embryonic development are quite different in their chemical nature from metabolic substrates which act in a plasmotropic manner in the induction of enzyme synthesis in bacteria.

The nature of truly genotropic substances still remains one of the greatest gaps in our understanding of the mechanism of differentiation. One of the questions which may be asked about them is the degree of specificity which we need to attribute to them, and from whence this might reasonably be supposed to arise. To fill the role we have assigned to them, the substances must be supposed to impinge on certain definite genes. But should we expect to find that each genotropic substance recognizes and acts on only one locus, or might it have various degrees of affinity with a number of different loci? The latter supposition might offer tempting possibilities of accounting for the activation of coordinated batteries of genes acting with an organized set of relative intensities, which seems to be demanded by the nature of differentiation, in which the formation of a particular type of cell, such as liver or muscle, must involve something of this sort.

Any substance specific enough to interact with only one or a few loci in the genome must, one would suppose, be a fairly direct gene product. If the specificity of a genotropic substance is not absolutely precise, it is not too difficult to imagine the substance which impinges on gene X being a product of gene Y. The interaction with X might, in such a case, be determined by some aspect of the tertiary structure of a protein, or some similarly complex property of an RNA molecule. Something of this sort is presumably envisaged for the genotropic repressor substance which is produced by the repressor gene and interacts specifically with the operator locus. The greater the specificity which we attribute to the interaction between repressor substance and operator gene; the greater the chemical similarity which we should have to suppose exists between them, and thus, the less the force of the objection raised previously to the whole theoretical system; namely, that it provides no grounds why a substrate should induce an enzyme capable of attacking it. But if we follow this line of thought, we soon reach the point of supposing that a substance capable of specifically reacting with only one locus in the genome must have been produced by that locus itself.

One recent investigation, which may prove illuminating about the nature of genotropic substances although its author does not interpret it in that way, has been reported by Goldstein (1958). If *Amoeba proteus* are fed on Tetrahymena which have been cultivated in a proteose-peptone solution containing added radioactive methionine, the amoebae become labeled quite heavily in the nucleus. A labeled nucleus can then be transferred by micro-surgery into the cytoplasm of a normal unlabeled Amoeba. If, after being fed on "hot" Tetrahymena, the donor Amoeba is kept for some hours in the presence of nonradioactive methionine before its nucleus is transferred, it is found that label passes from the grafted nucleus almost entirely into the nucleus of the host Amoeba, very little being found in the cytoplasm. This transfer might be a movement of nuclear proteins, but there is the possibility that the labeled proteins break down to liberate free amino acid, and that the label travels in this rather uninteresting form. But Goldstein showed that this possibility is unlikely; if the washing of the donor with label-free methionine is omitted, a grafted hot nucleus gives up label to the cytoplasm of the host as well as to its nucleus, which strongly suggests that free amino acid is taken up by both components, and not exclusively by the nucleus. Goldstein seems, therefore, to have demonstrated rather clearly a passage of protein out of one nucleus and into another. He suggests that this may be the protein component of a ribonucleoprotein, since he and Plaut (1955) had previously shown that label incorporated into RNA behaves in a similar way; and he seems to attribute greater importance to the RNA part of the hypothetical compound than to the protein. However, there is no evidence that the protein labeled by the methionine has any RNA attached to it; and Goldstein may really have been observing the synthesis in the nucleus of a protein which, after passing into the cytoplasm, can then go into another nucleus again as a genotropic substance might be expected to do. Similar experiments on cells in which intranuclear structures can be resolved in autoradiographs should prove very illuminating.

It is now time to consider another hypothesis, advanced from quite a different set of considerations, which issues in suggestions about the nature of genotropic substances. I shall choose to discuss one rather definite series of suggestions which have been made, not so much because they are yet fully proven, but because of their suggestiveness. This theory is the series of suggestions which has recently been developed by Leslie (1961). It is based in part on arithmetical manipulations of figures. Some people tend to remain rather sceptical in the face of such numerology, but although it is perhaps never entirely convincing when unsupported by anything else, it has often proved extremely suggestive. Was it not Astbury's alleged numerological identity, between lattice spacings he thought he had detected in proteins and nucleic acids, which opened up the whole nucleic acid protein story in the mid 1930s?

The main foundation of Leslie's theory is an analysis of the base composition of the DNA and various RNA preparations from a strain of human liver cells which were cultivated in tissue culture. DNA always exists in a duplex form, in which the bases which make up one strand are paired with those of the other strand; adenine (A) always being paired with thymine (T), and guanine (G) with cytosine (C). Thus the proportion of A is always equal to that of T, and of G to that of C. Leslie finds that in the human liver cells the proportions per hundred bases are 30A:30T : 20G:20C.

The cells were then fractionated by centrifugation into three portions: nuclei, particles (comprising nuclei, mitochrondria, and large microsomes), and ribosomes. When the base composition of the total RNA of the particle fraction was investigated, the proportions turned out to be 19A:18U : 31G:32C (uracil took the place of thymine). There is an approximate equality between A and U, and between G and C; and the rations are very nearly 20:20 : 30:30, the inverse of the DNA ratios. This strongly suggests that this RNA is duplex in constitution.

In other RNA fractions, however, the situation is different. If the particles are extracted with ice-cold dilute sulphuric acid, most of the histone protein and about half the RNA is removed, and the remaining RNA has a base composition of 14A:25U : 48G:13C. Again, from nuclei washed with citric acid, the total RNA has the base composition 24A:15U : 37G:24C. Thus in neither of these preparations is there anything approaching equality between the A and U, or the G and C. It is, therefore, very difficult to imagine the RNA being in duplex form; and Leslie interprets it as having only a single strand.

Since these RNA's which show asymmetry of bases and which are therefore presumably single-stranded, are derived from the nuclei and from a fraction containing many microsomes, it seems likely that they correspond to messenger and/or template RNA. If the base compositions given above are looked at more closely, a very suggestive fact emerges. Suppose the particle RNA with the composition 14A:25U : 48G:13C is a single-strand molecule, then add to it a hypothetical complementary strand, which would have to be 14U:25A : 48C:13G, and now sum the overall base composition of this duplex. We find that we have per two hundred bases, 39A:39U : 61G:61C, and this, reduced to give proportions per hundred bases, is very close to the composition of 20:20 : 30:30, which was found for the total RNA. Similarly, if a complementary strand is added to the supposedly single-stranded nuclear RNA, the total base composition of the resulting duplex comes out to almost the same set of ratios. Now there is no reason to suppose that the base composition of any particular gene must always be near the overall average composition. We should not expect to find this relationship if the single-strand RNA were derived from one or only a few genes. The fact that its base composition is such that, when a complementary strand is added to it, the total is restored to the general overall average composition, shows that the single-strand material must be derived from the whole, or at any rate from a sufficiently large sample of the DNA. And the fact that we have two different populations of RNA which both show this property indicates that in the derivation of the single-strand from the duplex material, sometimes one of the two strands is retained and sometimes another.

Leslie, therefore, draws the conclusion that the duplex DNA can potentially produce a corresponding duplex RNA, but that in the formation of the messenger and template RNA one of the two strands in this duplex is retained and the other is destroyed. This activity does not take place at only a few loci, but at a large enough proportion of them to reflect the overall base composition of the total DNA.

The next, and very important, step in the argument is the demonstration that histones, which are known to occur both in the chromosomes and in the microsomes, can act as ribonucleases. These histones, Leslie suggests, are highly specific in their actions, which may take the complementary forms of depolymerizing certain RNA's and combining into stable complexes with other RNA's. Leslie is, therefore, led to the hypothesis that the cellular histones may act to destroy one strand of a duplex RNA and simultaneously to stabilize the complementary strand at any locus. If, in the duplex RNA, only one of the two strands is an effective template for protein synthesis while the other is a "nonsense sequence," then histones which operate in this manner would determine which templates would be present in any given cell, and thus which gene-action systems would be operative.

Now there are obviously still many doubtful points in this theory. It is not obvious, for instance, why any given type of cell, such as the human liver cells studied by Leslie, should contain two different populations of single-stranded RNA. But again, the important point is that the theory is a worked-out example of the kind of consideration which is becoming actual in tackling the problem presented by the control of gene-action systems. Moreover, it draws attention to a factor which, it has long been obvious (cp. Waddington, 1956, p. 57), has been unduly neglected, namely, the participation of the proteins in the chromosomes and in the other synthetic systems in the cell.

We have still only rather inadequate techniques for studying cell proteins. Stedman and his co-workers (Mauritzen and Stedman, 1960), have shown by direct chemical analysis that certain histones are cell-specific, but these are presumably populations containing a number of different species of molecules; moreover, we know little about their location within the cell, although Leslie gives some reasons for supposing that they are intranuclear.

Several general considerations would lead one to expect that considerable changes are taking place in the nuclear proteins during the early stage of development when the control of gene-action systems must be actively in progress. For instance, the zygote begins life with the set of chromosomes derived from the father with an obviously unusual protein constitution, since in the sperm the DNA is combined in many forms with protamines instead of any of the more usual type of chromosomal protein; and in many species the other maternal set of chromosomes passes through some sort of lampbrush or germinal vesicle stage in which it seems likely that its protein composition is different from that characteristic of most adult nuclei. The simple morphological evidence strongly suggests that rather radical changes in nuclear composition take place in the very early stages of development. The breakdown of the germinal vesicle with the release of the nuclear sap and the appearance of contracted maternal chromosomes is one such indication. Further support for the same conclusion can be found in observations made by Edwards and Sirlin (1956) in our laboratory. Sperms from a normal male mouse were allowed to fertilize eggs which were heavily labeled in their cytoplasm following the injection of glycine into the mothers. Autoradiographs of sections of such eggs showed that, during the growth of the spermhead into the male pronucleus, label picked up from the cytoplasm had become incorporated into substances which resisted extraction during the histological processing, that is presumably into newly synthesized nuclear proteins.

It is, perhaps, worth pointing out that the histones, as normally encountered, are comparatively small proteins with molecular weights of the order of 3 to 8,000, that is to say, containing only some twenty to fifty amino acids. It would require a considerable number of them to cover the whole stretch of DNA in a normal old-fashioned gene. Is it possible that the subdivision of such a gene into a set of cistrons is a reflection of its coverage by a series of distinct histones?

Bloch and Hew (1960) have recently distinguished several of the basic nuclear proteins by means of their staining reactions with fast green, bromophenol blue, and other dyes. They showed that during the early development of the snail *Helix aspersa* there is a progressive change in the nuclear proteins. The sperms contain protamines (the spermatids having previously passed through a stage with characteristic arginine-rich histones). In the fertilized zygote, the sperm protamines rapidly disappear, and the cleavage nuclei show a cleavage histone, whose staining properties differ from those of adult cells. The changeover from the cleavage type to the adult type of histone occurs at about the time of gastrulation (see also, Alfert, 1958).

The suggestion, arising from Leslie's work, that histones may act not only as nucleases which destroy one strand of RNA but as compounds which stabilize the other strand, leads one's thoughts toward another area of biology in which the stabilization of one out of a number of possibilities is crucial. This is the field of immunology. There recently has been a considerably increased interest in immunology among embryologists, but in most cases the immunology has been used mainly as a method of preparing very specific reagents, which can be used to detect particular antigenic molecules in developing cells, as in the beautiful work of Holtzer, Clayton, Woerdemann, Langman and others (reviewed by Clayton, 1960). One wonders, however, whether it might not be worthwhile to consider the possibility of a closer parallel between the phenomena of embryonic development and the production of antibodies. In development, various regions of the egg cytoplasm and its derivatives control the gene-action systems in such a way that in a given area certain proteins are produced; in antibody production, some form of control by an administered antigen results in the production of a particular protein antibody. Can we suppose that there is any basic similarity in the two types of processes?

Most theories, both of development and of antibody production, operate with the concept of the gene as an entity which is at all times capable of transmitting the information necessary for the production of one and the same specific protein. The gene is thought to be not only invariant in composition, but also fully competent at all times to "code for" its corresponding protein. This raises difficulties in explaining how it comes about that the cells of a vertebrate body can produce an antibody protein whose structure corresponds to that of any given antigen which may be administered. One hypothesis would be that the structure of this protein is directly determined, at least in part, by the antigen; but this contradicts the basic assumption that the structure of proteins is determined by genes and only by genes; and if this assumption is abandoned, we should find our general theoretical structure in chaos. The alternative explanation usually offered is the "clonal selection" theory of Burnett (Burnett and Fenner, 1949; Burnett, 1959). According to this theory, the genome contains certain highly unstable loci which, in the antibody producing cells, mutate to produce a very large variety of alleles capable of controlling the production of a large variety of proteins. When an antigen is administered, this selectively encourages the multiplication of a clone derived from the particular cell which contains an allele capable of controlling the synthesis of the corresponding antibody protein.

This theory provides an adequate explanation of the phenomena as they are known at present, and also has the merit that it seems open to experimental testing. For instance, it would appear that the theory could hardly be rescued if individual cells were shown to be able to form antibodies simultaneously against *any* two antigens. However, the theory has one rather unsatisfactory aspect in that it attributes the production of antibody protein to a special type of process involving genes which undergo hypermutations into a vast range of alleles, a process which is not supposed to occur in the genes responsible for the normal tissue proteins.

If we wished to construct a theory according to which antibody production was seen as something more closely allied to normal development, we should, I think, have to picture the formation of tissue proteins as involving some stage in which the genes are more labile than we usually contemplate. One might, for instance, explore the consequences of supposing that all genes are, in the very earliest stages of development, both labile in the sequence of bases in the DNA (and therefore hypermutable), and at the same time not in a state in which they are capable of transmitting template information. We should then have to suppose that the genes become stabilized in their base sequence, and rendered effective transmitters, by combination with corresponding molecules derived from the cytoplasm (for instance, possibly histones). These cytoplasmic "co-genes" would have to be formed or modified under the influence of the maternal and paternal genes in the zygote. The suggestion, in fact, would be that the DNA-co-gene complex gradually become stabilized by a mutual influence of the two components; and the process of stabilization might go on at different rates, and even possibly reach different states in different regions of cytoplasm. Instances of cytoplasmic inheritance might be cases in which co-genes, formed in the egg cytoplasm, were somewhat resistant to the influence of paternal alleles. And the genes responding to injection of antigen by the production of antibody-protein would be those in which the stabilization of the complex proceeds very slowly, and is only completed by the action of the antigen itself. The induction of immunological tolerance, or the failure to produce antibody against self-proteins, would have to be considered akin to immunological paralysis, produced by a precocious stabilization following the early administration of massive doses of antigen.

Again, this hypothesis is not offered as something which is necessarily convincing as it stands. It is put forward to provide another example of one of the patterns of thought that are beginning to open new avenues for our exploration. Although genes obviously have sufficient stability to be reliably transmitted through many generations, the occurrence of immunological phenomena which seem to demand the existence of hypermutation suggests that we need not feel ourselves bound to regard genes as nothing more than firmly bounded sequences of DNA, modifiable in no other way than by rare events of mutation.

The Cytoplasmic Elaboration of Substances

So far we have discussed only the control of that part of a gene-action system which leads from the gene to the first protein whose amino acid sequence the gene determines. It is obvious without further discussion that many of the phenotypic characters by which the activity of a gene is recognized, such as various pigments, arise as the result of further metabolic processes taking place between and among the primary gene-produced proteins. It is, perhaps, not quite so obvious, but still seems to be the case, that this is also true of many of the proteins in the form in which they occur in living cells.

The simplest examples of these proteins, which may be called "secondary," are those which normally exist as polymers, in which two or more simple protein molecules are coupled together. Hemoglobin, for instance, is a double dimer, consisting of a pair of molecules of one type coupled with another pair of a different type. In normal adult hemoglobin there are a pair of α chains coupled to a pair of β chains. These two types of chain are manufactured by two different loci, each of which can mutate to a number of alleles (for recent reviews, see Ingram, 1959; Anfinson, 1959). The normal allele found in most adults is called the A allele at both loci, so that constitution of normal adult hemoglobin can be written as $\alpha_2^A \beta_2^A$. During fetal life, another locus, γ, is in operation while the beta locus is inactive, fetal hemoglobin thus having the composition $\alpha_2^A \gamma_2^A$.

There are still many interesting points we do not know, but soon may discover about the workings of these gene-action systems. For instance the alpha and beta chains must be rather closely correlated in structure, so that they fit together and leave room for the large heme groups. Are they manufactured in the same or in different microsomal particles? Clayton, in our laboratory, is developing methods for labeling antibodies with substances which can be recognized in the electron microscope; and with these it may be possible to answer questions of this kind.

One point that is clear, is that the form of the final molecule is not determined fully by the structural genes, but arises as a secondary coupling of the individual chains. This is shown by a case recently described by Baglioni and Ingram (1961). A person was found with four hemoglobin types: hemoglobin C, $\alpha_2^A \beta_2^C$; hemoglobin A, $\alpha_2^A \beta_2^A$; hemoglobin X, $\alpha_2^G \beta_2^C$; and hemoglobin G, $\alpha_2^G \beta_2^A$. From this it can be deduced that this individual had the genotype $\alpha^A/\alpha^G \beta^C/\beta^A$. The alpha and beta chains always associate in dimers consisting of two completely similar partners; we never get an α^A going with an α^G. But the two types of alpha dimer can be coupled at random with the two types of beta dimer.

Another phenomenon which probably finds its explanation in interactions between primary gene-produced proteins is that of allele complementation. This process occurs in several organisms, but has been particularly studied in Neurospora; a recent, and good, account will be found in Catcheside (1960). In this paper, Catcheside describes results of a study of 389 independant mutations (induced by ultraviolet) affecting the synthesis of histidine in *N. crassa*. These were found to belong to six loci, known as *his–1, his–2, his–3, his–5, his–6* and *his–7*. The missing locus in this series, *his–4*, is known as a single mutant described in earlier work; its growth characteristics are such that it would probably not be picked by the particular technique of searching for mutants which was used in this study; therefore, its absence from the material to be discussed is not surprising or significant.

The six loci described affect various steps in the biosynthesis of histidine, as shown in Figure 5. Complementation arises when two alleles of

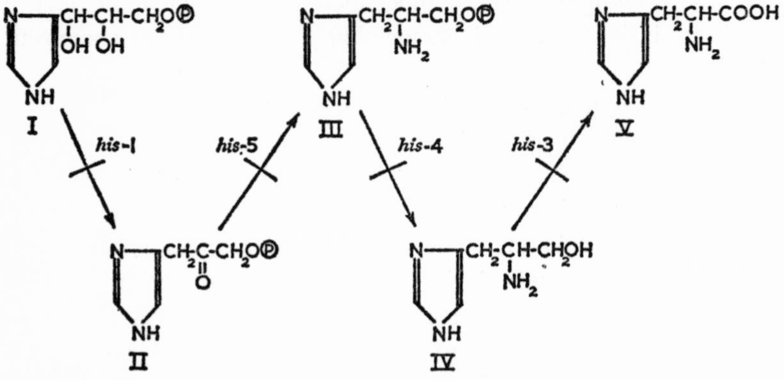

Figure 5. Pathways of histidine synthesis in Neurospora crassa

The effects of four of the known alleles are indicated; the other alleles act on synthetic processes leading up to substance I. (From Catcheside, 1960)

a single locus brought together into the same multinucleate heterokaryon allow the missing enzyme to be produced and the previously inhibited step in the synthesis to proceed. Phenomena of this kind were found to occur among alleles of four of the loci, but not in the other two. For one of these two, the testing for the occurrence of complementation was very extensive, some 95 alleles being tried together in all the 4,465 pair-wise combinations. It, therefore, seems rather likely that complementation really does not happen at this locus (which was *his–6*).

At the loci where complementation does occur, it happens only between certain pairs of alleles. The relationships can be expressed in diagrammatic form in two different ways (Figure 6). The alleles of the locus will be found to fall into a certain number of groups which may be denoted by the letters A, B, C, D. In the complementation matrix form of diagram, these letters are written horizontally and vertically, and at the intersection between, say, the B column and the C row an open circle is written if no complementation occurs in this combination of alleles; a full circle if it does occur. In the complementation map type of diagram, each group of alleles is represented as a line, and these are placed in relation to one another so that those for groups among which complementation occurs do not overlap. It is a surprising fact that in nearly all the loci studied it was possible to arrange the complementation map in such a way that each group of alleles was represented by a single line; only in the case of *his–1* does the map have to take a form in which one group is represented by two separated lines (alternatively the map in this locus could be represented by a set of two-dimensional diagrams).

These maps are originally derived as purely formal representations of

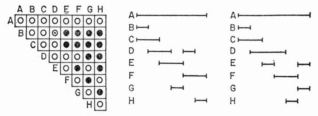

Figure 6. Complementation between alleles at locus histidine-1

On the left, the relations are shown as a complementation matrix; a black circle at an intersection between two alleles indicates that they complement one another, a white circle that they do not, and a circle with dot that they complement slightly. On the right, the same relations are indicated in two ways by linear diagrams; two alleles complement only if represented by lines which do not overlap. This locus is the only one of those studied in which such a diagram requires one or more alleles to be represented by two separate lines (D on one system of representation, E on the other). (From Catcheside, 1960)

complementation relations, which are equally well described by the matrices. We are, however, so used to the reality of linear arrangements in genes that it is natural to enquire whether the complementation maps also are a direct representation of some material structure. Such an idea would be quite plausible for a locus in which there were only three groups of alleles, one showing no complementation and the other two complementing with each other. We should then have a complementation matrix and map as in Figure 7, and this could be explained by supposing that the gene had a tripartite structure, for example, an arrangement of regions in the order B A C, where B A is involved in carrying out one function and A C with carrying out some other function, and both functions being necessary for gene activity. Then in a heterokaryon containing B′ A C in one nucleus (the prime indicating a mutation) and B A C′ in the other, complementation would occur, since one nucleus would be uninjured in relation to one function and the second nucleus uninjured in relation to the other; nuclei carrying mutants in A would be injured in both functions and could not be complemented by any other single nucleus.

However, this explanation becomes less and less convincing the larger the number of complementation groups. If there are three groups B, C, and D, showing complementation between them, it is more difficult

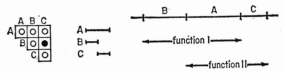

Figure 7. Diagram and interpretation of simple complementation series of three alleles

On the left is a simple complementation series of three alleles, represented by matrix and linear diagrams. On the right is the interpretation of the linear diagram as a representation of a sequence of functional sites along the chromosome. Allele A, which lacks both functions I and II, would not complement with either B or C, but these would complement with each other. (From Catcheside, 1960)

with schemes of this type to see how there can be a fourth group A which shows no complementation with any of the others. If we wish to suppose that the A mutants affect a region of the gene which is common to B, C, and D, we have either to suppose that these mutants always involve deficiencies which extend over longish regions of the gene, or that they affect two or more separated sites in the gene. Both these suppositions would lead us to expect the noncomplementing mutants fail to backmutate to normality, or to do so only in extremely low frequency; whereas, in fact, they are found to backmutate at a normal rate. It seems, therefore, that we have to regard the complementation maps as purely formal and to look for some other type of explanation for the phenomenon.

One variety of hypothesis attributes complementation to some process akin to crossing over. It might be, for instance, that the genes in the two nuclei produce two differently defective template RNA's, and that these can recombine with one another to reconstitute a normal template; or the recombination might be supposed to go on between the amino acid chains of the defective proteins. In all such recombination theories, however, it is difficult to explain why, if complementation is possible at all, it should ever fail. Thus without a lot of subsidiary hypotheses—and it is not clear what form they should take—no explanation is offered to the occurrence of noncomplementing alleles and of complementation groups.

The type of hypothesis which is favored by Catcheside is the supposition that complementation depends on the formation of partially effective polymeric complexes between two or more elementary proteins each of which is defective. This could, of course, only happen in relation to a gene whose final protein product exists as a polymer; thus, there is a natural explanation for the complete failure to discover complementation in some loci, such as the *his–6* locus mentioned above. Moreover, since the activity of a protein seems usually to depend on the precise stereochemical arrangement of some relatively small part or parts of it, it is not too difficult to see, in general terms, though not yet in detail in any case, how differently defective protein chains might complement one another in certain combinations in building up at least partially effective polymers. The level of enzyme activity in the complemented heterokaryons in the few cases in which it has been determined (Fincham, 1959) is very low. The hypothesis of complementation by polymer formation is, in fact, although not yet a very precise one, at least powerful and flexible. It can certainly suffice to draw our attention to another new pattern of thought concerning the control of gene-action systems.

Another example of the control of a gene-product which is, perhaps, best envisaged in terms of secondary reactions between proteins, is a story which seems at first sight to upset a number of theoretical applecarts. Most proteins are made up from only twenty different amino acids; and those who have thought about the correspondence between nucleotide sequences in DNA and the amino acid sequences in protein have been able to work out certain more or less plausible codes, which would make it possible to specify just twenty amino acids and no more. It is rather upsetting to such theories to find that there are some proteins which contain unusual amino acids over and above the twenty which form the normal range.

One such protein is the flagellin which can be extracted from the flagellae of Salmonella; in certain strains of bacteria this contains ϵ-N-methyl-lycine. Some Salmonellas occur in two phases, and in these the flagella antigens of the two phases are determined by two different genes,

H_1 and H_2. In a diphasic strain carrying the gene for N-methyl-lycine, both the two phases contain this amino acid in their flagella proteins. This makes it seem probable that the presence of this amino acid is due to some gene other than the H_1 and H_2 which are responsible for the sequences of the normal flagella proteins. In transduction experiments, Stocker, McDonaugh, and Ambler (1961) found evidence that the presence of this amino acid is dependent on a gene which is certainly distinct from H_2, and is probably distinct from H_1, although closely linked to it. The probable explanation of the situation seems to be, then, not that we are confronted with a case in which DNA directly codes for an amino acid outside the normal twenty, but rather that the N-methyl-lycine gene produces an enzyme which secondarily modifies the H_1 and H_2 proteins by causing the substitution of this acid for one of the amino acids which they already contain.

This raises the possibility that such secondary rebuilding of proteins may be relatively common. If it involved the substitution of one of the normal twenty amino acids for another, it would be very difficult to detect; in fact it could probably only be demonstrated when we know enough about DNA and its code to make direct comparisons between certain genes and their corresponding primary or secondarily modified proteins.

Conclusion

We have reviewed in this chapter a whole series of cases which suggest new patterns of thought in relation to the basic problem of embryological development: which is the problem of which genes shall be active, in a dominating manner, in which cells. We have had to consider the possibility of genes exerting direct control on other genes, as Monod and Jacob's Operator does on its set of Operon genes; of genes influencing other genes through a nonchromosomal substance, as the Regulators do on the Operators; of influences on the messenger or template RNA's which intervene between the gene DNA and the primary proteins; and finally of a potentially flexible system of interactions between primary proteins, leading to the formation of polymeric secondary proteins and eventually to grossly phenotypic characters. In thinking about the problem of how one cell becomes a muscle cell which synthesizes myosin, actin, etc., and another a pancreas cell which produces insulin and various other things, a whole spectrum of new, or at least renovated, lines of thought have been presented. In particular, we are being forced to make serious attempts to identify and characterize genotropic substances.

But to the embryologist who likes to preserve a moderately hard-headed attitude, these thoughts seem stimulating but a bit in the air. We want to bring into the picture another obtrusive fact: cell activities tend to be organized. A cell which produces myosin tends to produce actin too. And we also have been conditioned by our upbringing on a diet of microscopy, to like to see our theoretical concepts as something we could put our fingers on, if only they were small enough. These two problems, of the integration of gene-action systems into organized complexes, and what developing cells actually look like, will be the topics of the next chapter.

2. Kinetic Organization and Cellular Ultrastructure

*I*N the first chapter were discussed the new patterns of thought which are arising about the nature of individual gene-action systems, and the ways in which they can be controlled, either by influences which act on the genes themselves, or by effects on the gene-protein systems, or on the secondary reactions which intervene between the primary proteins and the phenotype. I shall now go on to consider another level of complexity. It is inadequate to envisage the histogenesis of metazoan cells simply in terms of the activity or nonactivity of individual gene-action systems.

This is so for two main reasons. In the first place, there are several grounds for believing that, in any particular cell type or tissue, many more gene-action systems are in operation than are apparent at first. One good argument for this suggestion is the old evidence of Demerec (1936), who found that a surprising proportion of an arbitrary collection of small deficiencies acted as cell lethals in the hypodermis of Drosophila. In the second place, zygotes whose genotype has been molded by natural selection commonly exhibit the phenomenon which I have called the "canalization of development," that is to say, they can carry out histogenesis into a restricted number of alternative end states among which there are few if any intermediates. Each such pathway of developmental change, which I have called a "creode," must involve the operation of a considerable number of gene-action systems, and these are interrelated by some type of feedback connection in such a way that the developing system, if diverted to a minor extent from the creode, has a tendency to return to it.

I have recently discussed these matters at some length (Waddington,

1956, 1957, 1961a). For our present purposes the main point is that any general theory of development should envisage each pathway of histogenesis, or creode, as the resultant of essentially all the available gene-action systems whose intensities are mutually adjusted by interlocking control systems. However, this does not deny that in certain tissues the activity of some particular gene-action systems may be nil; or even that certain genes may, in some tissues, be completely inactivated or lost; but these should be regarded as extreme cases of a more general control over the intensity of operation of every gene-action system contained in the cell. We have to consider the organization of the kinetics of gene activities in the cell in terms which are more flexible than a simple on-off switching.

Toward a Mathematical Theory of Epigenesis

In this chapter, I will discuss some new patterns of thought in relation to the cell as an organized entity. These new patterns concern particularly what one might think of as the two extreme ends of the spectrum of possible approaches to this problem. At the one end, is some extremely abstract theory which attempts to set up a mathematical model of the general kinetics of cell synthetic processes; at the other, is a procedure which is very simple in approach but nowadays has new sophistication in method, that is examining actual cells to see what they look like.

Biology in general has been very weak in the development of theories of wide application. The subject which is usually known as "theoretical biology" often turns out to be dealing with questions that are, at least, partly philosophical in character, such as the vitalist-mechanist controversy. In recent years there has been a movement, led by Woodger (1937), to develop a theoretical biology in terms of logical analysis; but on the whole the problems which can be dealt with in this way are simple ones, such as the relations among entities which form part of a hierarchical system based on various kinds of one-to-two relationships; and these are not problems crucial to our understanding of biological processes. What we should, I think, like to possess is a body of theory comparable to the major physical theories, such as thermodynamics, general relativity, wave mechanics. Population genetics and evolution are almost the only divisions of biology for which theories of this character are available.

In histogenesis, the general terms in which a theory has to be framed are fairly clear. We are dealing with sets of enzymatically controlled synthetic processes, which interact with one another because they utilize the same basic raw materials, amino acids, purine and pyrimidine bases, sugars, phosphate, etc., and because the product of one synthesis may act as a specific catalyst or inhibitor of another. One might have hoped that chemical engineers, dealing with flow systems in continuous processing plants, would have elaborated a theory which we could apply to embryos; but if they have done so, I have been unable to discover it. The original notion that such an interlocked system of syntheses could be so adjusted by natural selection that it issues in a set of alternative creodes was advanced on almost purely intuitive grounds (Waddington, 1940a). Later, the first tentative approaches toward a mathematical theory were made (Waddington, 1954) by the application of equations which have been considered by demographers, in particular by Volterra (1931), in connection with the dynamics of population growth. Catalytic synthetic processes which compete for the same substrates and which specifically enhance or inhibit one another, have the same formal properties as populations of different species of animals which compete for the same nutritive materials and may also be related to one another as predator and prey.

Unfortunately the Volterra equations, in their most general form, are extremely difficult to handle mathematically, and until recently little progress has been made in developing from them any important theoretical consequences. However, the demographer Kerner (1957, 1959) has made important progress in generalizing the equations in a form which can be handled mathematically, and his results have been adapted to the situations arising in development by the mathematician Goodwin (1961) working in our laboratory.

In development, we are confronted—as genetics has brought home to us—with extremely complex systems involving very many components. Presumably one exists for each of the genes which affects the particular creode with which we are concerned. In such circumstances, the first question to ask is not, what is the detailed nature of the components— we can hardly expect to get an answer to this—but rather, what stable states will survive a given disturbing stimulus? That is to say, what we need is statistical mechanics comparable to the thermodynamics of physical theory. But we shall have to elaborate one that applies to open systems which do not conserve either matter or energy; whose final state is not determined by the initial conditions; and in which entropy can decrease. Goodwin claims to have developed, from somewhat modified Volterra equations, concepts which can play theoretical roles analogous to those taken by such notions as total energy, kinetic energy, temperature in physical theory. No names are available for these concepts in relation to development, and they are referred to by the most suitable physical name qualified by the adjective "epigenetic"; so we speak of the "epigenetic temperature," "epigenetic kinetic energy," etc. I shall give just the bare outline of the mathematical development, partly to convey some inkling of the nature of these concepts, and partly to hope that other mathematicians may be stimulated to carry the work further.

Consider a system with many interacting autosynthetic components X_1, X_2, X_3, etc. Then for the ith component X_i

$$\frac{dX_i}{dt} = X_i[\alpha_i(X_1, X_2 \ldots X_n; a_1, a_2, a_3 \ldots a_n)],$$

where the a's are environmental factors. A steady state will occur when there is no further change in X_1, that is when the expression on the right is equal to zero.

The simplest case, and that considered first by Volterra for animal populations and myself for autosynthetic chemical species during development, is to suppose that the components are competing with one another for substrates. The equations then become simplified to

$$\frac{dX_i}{dt} = X_i[\alpha_i(a) - f(X; a)],$$

where α_i is different for each component, but $f(X; a)$ is the same for all. Such a system is unstable in the sense that one component always runs away and the others disappear, so that in the steady states there is only one component left.

Williamson (1957), considering animal populations, tried to produce a system of equations that would give steady states in which several components persisted; and in order to do this, he introduced controlling factors or preference factors g. These factors, which can be interpreted in various ways, are monotonically increasing functions, such that $g_i(X_i, a)$ is positive when X_i is small, but becomes negative when it is large. The equations including these functions are

$$\frac{dX_i}{dt} = X_i[\alpha_i(a) - g_i(X_i; a)f(X_i; a)].$$

These give rise to a set of steady states, each of which is surrounded by a region over which it dominates, that is, states of the system within a given region will converge on the corresponding steady state. However, without further knowledge of the character of the functions g and f, no general integral can be found for such a set of equations, and we can make no progress toward using them to develop a statistical mechanics.

A particular way of dealing with the interactions between components which is favorable for further analysis is to treat them as always representable as the sum of all possible binary interactions between components taken two at a time. This is, of course, a restriction on the generality of treatment, but perhaps a legitimate one, at least for a start. The equation then becomes

(1) $$\frac{dX_i}{dt} = X_i\left[\alpha_{io} + \frac{1}{\beta_i}\Sigma_j \alpha_{ij}X_j\right],$$

where $\alpha_{io} = \alpha_i(a)$ is a function of the external parameters a, and β_i is a factor to take account of the possibility that an interaction between two components may involve a number of units of X_i interacting with one unit of X_j. The α_{ij} may be any function of the X_i variables, but for further developments must have the antisymmetric property that $\alpha_{ij} = -\alpha_{ji}$ so that $\alpha_{ii} = 0$, and there are no logistic terms in the squares of the components, such as X_i^2. If we call the stationary state value of X_i as q_i, these will be given by the equations

$$\alpha_{io}\beta_i + \Sigma_j \alpha_{ij}q_j = 0.$$

Kerner showed that if we introduce a new variable $x_i = \log\dfrac{X_i}{q_i}$, we get a solution of equation (1) above in the form

$$\sum_{i=1}^{n} \beta_i q_i(e^{x_1} - x_1) \equiv \text{constant} \equiv G \equiv \Sigma_i G_i.$$

This integral G corresponds to the total energy in physical theory, and it can be used as a basis for a statistical thermodynamics in a closely analogous way.

Goodwin derived, for the context of development, two concepts which may be called "epigenetic temperature" and "epigenetic kinetic energy." The former, which is defined as

$$\theta = \frac{\beta_i - q_i \overline{(x_i - q_i)^2}}{q_i^2},$$

can be roughly interpreted as the mean square deviation of the concentrations of the components from their steady state values. It is thus, in thermodynamic terms, a measure of the degree of excitation of the system. Goodwin provides reasons why one might expect it to be high in early stages of development and gradually to fall. The epigenetic kinetic energy, defined as

$$T_i = x_i \frac{\partial G}{\partial x_i} = \beta_i q_i \left(\frac{x_i}{q_i} - 1 \right) \log \frac{x_i}{q_i},$$

is less easy to translate into terms appreciable by the mere biologist.

However, the extremely important theoretical result emerged that the components in a system may be expected at any given time to fall into two more or less discontinuously separate categories, according as their epigenetic kinetic energy is less than or greater than their epigenetic temperature. The latter will be major components, and will, as time passes, remain in fairly high concentration, fluctuating to a certain extent around their stationary value. The former are minor components, usually present only in very small concentrations, but occasionally showing a burst of activity which brings them into the region of, or even temporarily above, their stationary value. The discontinuity between the major and minor components is inherent in the nonlinearity of the initial equations. Its importance is that it provides a very general way of accounting for the possibility of switching the system from one state in which certain components are major (i.e., one type of differentiation is under way), to another state in which other components have become major ones.

One can, in fact, obtain a theoretical relation giving the amount of epigenetic work required to raise a component from minor to major status, that is, to carry out a process of induction; and one can express mathematically what is meant by such embryological terms as competence (T not much less than θ), progressive determination (the difference $\theta - T$ gradually rising), and so on.

Now I should be the first to admit that all this theory is, for most biologists at any rate, completely "up in the air." It deals in terms of the fluctuations of the concentration of particular protein species; and we are not yet in a position to measure such concentrations. (The period of such fluctuations, Goodwin argues, may be of the order of ten seconds, while the comparable times for systems usually dealt within physical theory is some 10^8 times less. Most embryological processes occupy intervals which are sufficiently greater than ten seconds for the application of improved methods of protein estimation, which would allow the theory to be tested, to be not out of the question.)

But even before it becomes a practical proposition to test the theory, it is, in my opinion, of the greatest importance that theoretical developments of this kind should take place. I would remind you that the mathematico-genetical theory of natural selection is largely inapplicable in practice, in that it is expressed in terms of variables, such as effective population number, which are almost impossible to measure. But no one could possibly deny the enormous importance which these theoretical developments have had in providing a general framework of concepts within which our thoughts about evolutionary processes can move without so much danger of drifting off into mere cloudy verbalizations. What I think, and hope, we may be seeing in these new types of mathematical thought about epigenetic problems is the first tentative groping toward a really flexible theory; beginnings comparable perhaps to the integral equations relating nonoverlapping generations of infinite size, in which single genes were segregating, with which Haldane in the Proceedings of the Cambridge Philosophical Society in the 1920s, initiated the whole development of mathematical evolutionary theory, which so soon, in the hands of Wright and Fisher, overcame many of the limitations of its first over-simplified formulations. Goodwin's epigenetic thermodynamics possibly does not get us very far; but at least we have started on what is certainly a very novel pattern of thought; and one which I am sure will not stay put where it is now.

Cytological Structures

In the last decade or so, we have acquired two new methods of observation: electron microscopy, which allows us to see structures much smaller than anything visible with the light microscope, and autoradiography and other tracer methods which make it possible in effect to observe metabolic activities.

The logical place at which to begin a survey of the cell is with its most important element, the nucleus, and in particular the chromosomes. Unfortunately no one has yet discovered a satisfactory technique for fixing chromosomes so as to reveal much detailed structure within them in the electron microscope, so this new technique has as yet contributed little of major importance to our knowledge of these organelles. However, as some compensation for this, many extremely illuminating new observations have been made with the light microscope on various forms of giant chromosomes.

The most fully studied giant chromosomes are, of course, the salivary gland chromosomes of Drosophila and other Diptera. As is well known, Beerman, Mechelke, Pavan and their collaborators have discovered various species of Rhynchosciara, Chironomus, and other genera in which polytene or "salivary-type" chromosomes are well developed in a number of different tissues. They have been able to establish the points, which were to be expected *a priori* on the types of embryological theory we have been discussing, that different gene loci show signs of activity in different tissues, and further, at different developmental stages within the same tissue.

The manifestation of activity takes the form of a swelling of a particular locus on a chromosome, so that a region which, when inactive appears as a sharply demarcated band, becomes converted into a "puff" or "Balbiani-ring." When the phase of activity comes to an end, the puffed region regresses more or less completely to its previous condition, although in some cases it never becomes restored completely to the appearance it had before the puffing occurred. For instance, Pavan (1958, 1959), Rudkin and Corlette (1957) found that in Rhynchosciara some bands become very heavily Feulgen staining just before puffing but regress completely to their original state after the puff phase is over; others regress only to the heavily staining condition; while still other bands do not stain particularly heavily before puffing but do so after (Figure 8). Such increased Feulgen staining does not seem to occur in Chironomids (Beerman 1956, 1959a, b). It appears to involve changes in the DNA content of the bands in question; presumably this must be nongenetic DNA, not actually forming part of the information-carrying structure of

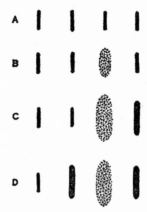

Figure 8. Four types of band behavior in Rhynchosciara salivary chromosomes

A, no change; B, formation of puff which regresses; C, formation of puff which regresses to a band containing increased amount of DNA; D, increase of DNA before and after puffing. (From Pavan, 1959)

the gene. The existence of such material would not be expected on present genetic theory.

There is a fairly gradual transition in morphology between the largest Balbiani-rings, through large puffs, to small puffs, to diffuse bands, and thus to the typical well-demarcated bands. This transition warns us, if nothing else does, against assuming that in any given type of cell, it is only the puffed bands which are active. It is safer to assume that all bands may be active to a greater or lesser extent; the puffed regions being those in which the production of material by the locus in question is too rapid to allow diffusion or other processes to remove it, so that some accumulation occurs. It would seem quite possible that there may be genes whose activity, even at its most intense, would never be manifested as puffing, because their product was so readily diffusible that it never accumulates. What we learn from the occurrence of different puffing patterns in different tissues and at different phases of development is that developmental change involves alterations in the intensity of activity of various gene-action systems; but we do not learn exactly what these alterations are.

An important step in the understanding of the puffing phenomenon will be the ability to control it experimentally, and some work on this

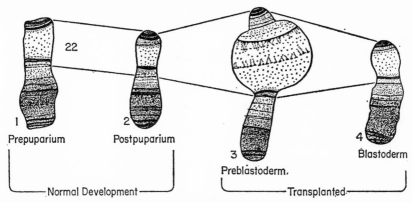

Figure 9. Banding pattern in chromosome II R of Drosophila bucksii

The two drawings on the left show the development of the banding pattern of chromosome II R of *Drosophila bucksii* around the time of puparium formation. The drawings on the right show the effect three hours after transplanting a prepuparial salivary nucleus, cleaned of most cytoplasm, into the cytoplasm of either a preblastoderm or blastoderm *D. melanogaster* egg. The cytoplasm of the pre-blastoderm egg induces the formation of a strong puff by region 22. (From Kroeger, 1960)

line is already proceeding. The ideal experiment would be, perhaps, to transfer a chromosome (or even a whole nucleus) from one type of cell to another and to show that it then acquires the puffing pattern characteristic of the host cell. This would be a complete demonstration that the puffing is induced by the cytoplasm in the manner implied by the cytoplasm-gene feedback loop discussed in the last chapter.

This experiment has not yet been done, but an approximation to it was carried out by Kroeger (1960). He removed the nucleus from a *Drosophila bucksii* salivary gland cell and placed it in cytoplasm squeezed out of eggs of *D. melanogaster* and demonstrated a characteristic alteration in the puffing pattern. This was not simply a degeneration phenomenon, since a change which occurred when the cytoplasm was from a pre-blastoderm egg did not happen in older cytoplasm (Figure 9).

Becker (1959), however, has described rather different results of isolation of the whole glands in saline. He first worked out the sequence of puffings at the end of the left arm of the 3rd chromosome (Figure 10). When a salivary gland is isolated in Ringer solution, its chromosomes may behave in three different ways. If the isolation is made at a comparatively early stage, while the puffs, which normally reach full development at about the time of pupation, are still in process of their early growth, then the sequence of puff changes becomes reversed, and the chromosomes eventually get back to the condition characteristic of the larva. If the isolation is made somewhat later, then the sequence of puff changes is carried forward, and the chromosomes come to resemble normal prepupal chromosomes. If the isolation occurs at the critical point, some nuclei in the gland go forward while other nuclei go backward, and they all develop an unusual puff on a band which never shows this activity in any other circumstances.

It will be noticed in Figure 10, and this point has been frequently made by all those working on these phenomena, that in all species studied there are many important puffs which develop at the time of pupation. The most direct experimental evidence yet available, in which puffing can be controlled by changing the environment of the chromosomes, comes from the work of Clever and Karlson (1961), who showed that when the pupation hormone ecdysone is injected into Chironomus larvae, the puffing sequence characteristic of pupation is brought into action. One of the very first responses to the hormone which can be detected in the larva is the development of a puff in section 18 of the right arm of chromosome 1 and the disappearance of a larval puff in section 19; and this is followed by the puffing at a number of other loci. It does not necessarily follow that ecdysone actually reaches the position of these regions of the chromosomes, and that it is to this actual substance (whose chemical formula is completely known) that the gene locus reacts by puffing. It would be very significant if we could be sure that ecdysone is really genotropic, but at present all that we are entitled to believe is that the ecdysone alters the cell in some way to which the gene locus responds by puffing.

Another point, which is illustrated by Figures 8 and 10, is that puffs tend to be transient phenomena. It is not often observed that a band remains in a puffed condition for very long; at any time when there is a

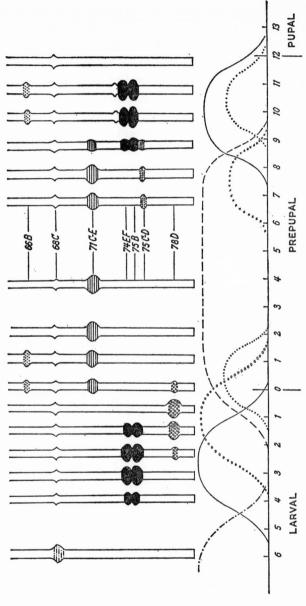

Figure 10. Puff changes in chromosome arm III L of Drosophila melanogaster

Puparium formation occurs at time 0; numbers to the left are hours before, those to the right are hours after. The dotted, dashed, and solid curves indicate the behavior of corresponding puffs. (After Becker, 1959)

physiological change of state in the cells, a number of puffs appear and then disappear again. Moreover, puffing seems to be something of an all-or-nothing process; if a band becomes puffed, it usually puffs up to its full capacity (Mechelke, 1961). Both these points suggest that the genic activity which we see in puffing is concerned rather with initiating specific protein-synthesizing systems rather than with keeping them going once they have started. One can, perhaps, find here some support for the suggestion, which I shall discuss later, that any one gene may only have to act intensively during a short period of development, and that thereafter the cytoplasmic entities which have received "instructions" from it can carry on more or less independently.

Another type of giant chromosome which has recently been studied with great profit—particularly by Callan (1955), Callan and Lloyd (1960a, b), Gall (1958), and see Waddington (1959)—is the lampbrush type found in the germinal vesicle nuclei of the oöcytes of many vertebrates. They are best seen in the oöcytes of newts. The germinal vesicle contains twelve bivalent chromosomes, which exhibit both chiasmata and also certain secondary fusions between some of the chromosomal products which will be mentioned later. In the present context we are not directly concerned with the rather surprising fact that the axis of each chromosome consists, not of two chromatids as classic theory would lead us to expect, but of a single very fine filament. This bears a row of small dense Feulgen positive chromomeres. The interchromomeric thread cannot be seen to stain in Feulgen, but is dissolved by DNAse. From each chromomere, a pair of loops (or a number of pairs) extends laterally to the axis. In fixed preparations, these are Feulgen-negative, and give evidence of containing protein and RNA; however, if the matrix which forms the bulk of the loop material is dispersed by saline and the loops are carefully dried on electron microscope grids, it can be shown that each loop contains an axial filament. These filaments appear to be exactly similar to the chromonemata, and like them are destroyed by DNAse. It seems to be the case, then, than the loop axis is actually part of the chromonema.

Around the axes of the loops there is a quantity of matrix. This takes a variety of forms and also varies greatly in quantity from locus to locus (Figure 11). It presumably represents the product of the particular gene at that position. There are a number of loci with giant loops, and

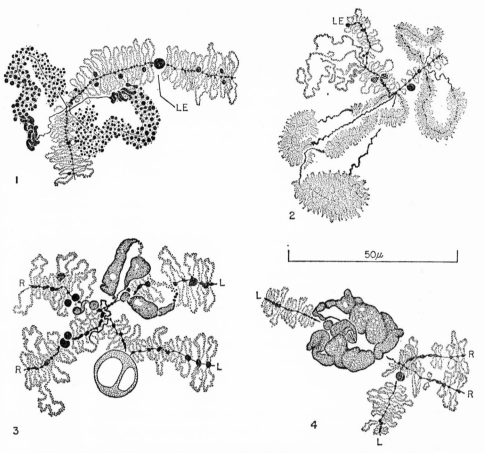

Figure 11. Small parts of some lampbrush chromosomes of crested newt
Drawings 1 and 2 are the left arm terminal regions of chromosome XII. In 1, from *Triturus cristatus cristatus,* the terminal granules of two partner chromosomes are fused at LE, and one chromomere bears two large coarsely granular loops. In 2, from *T.c. karelinii,* are two finely granular loops which are not in the same location as the coarse loop in other subspecies. Figures 3 and 4 are two examples of heterozygosity. In 3, from *T.c carnifex* bivalent XI, a pair of giant loops are fused in only one chromosome. In 4, from bivalent XII of same subspecies, a giant loop is present in only one chromosome. (The scale applies for 1 and 2, but is only approximate for 3 and 4.) (From Callan and Lloyd, 1960a)

a more or less complete gradation down to chromomeres with only very small loops; the centromeres, and in some species a number of chromomeres on each side of them, have no loops. In general one must conclude that every locus in the genotype is active to some extent, but that some are very much more active than others in this particular type of cell.

The most direct evidence that loop material represents the product of gene activity is the fact that, although most loops have exactly the same form at the corresponding loci on the two paired chromosomes, several cases are known in which they differ (Figure 11). These are to be interpreted as cases of gene heterozygosity, which here comes to visible expression in the chromosomes themselves. Such breeding results as are yet available fully support this interpretation.

The distribution of matrix in a loop is asymmetrical, one end of the loop being thicker than the other; the direction of asymmetry is constant along a single chromosome arm. Callan suggests two possible explanations for this asymmetry. According to one, the loop axis contains a series of genes whose activities fit together into a sequence like that of a flow production factory in which the thin end of the loop corresponds to where the operations start, and the thick end to where the sequential processes of synthesis reach their final stage, when the material would be ready to be shed from the gene locus. Although such "assembly lines of genes" are known in bacteria and some other lower organisms, they have never been certainly detected in higher organisms, and some special hypotheses would certainly be necessary to make it plausible to suggest that all the loci in newts are organized in this way. Moreover, as Callan points out, the matrix products are not always shed only from the thick end of the loop, but may come off at any point along it. For these and other reasons he prefers his alternative hypothesis: that each loop axis produces only one kind of product, the specifying structure for this product being repeated several times along the length of the axis. The asymmetry of matrix thickness is then due to the fact that the loop axis is pushed out from one half-chromomere and after passing some time in the extended and active condition, eventually taken up again by the other half-chromomere (Figure 12).

This hypothesis provides an explanation for the observations, but depends on the somewhat unexpected suggestion that the genetic information at each locus is repeated a number of times along the length of the chromosome. It seems, indeed, very difficult to avoid this conclusion. The length of the longest loops in the newt oöcytes is of the order of 60 μ, that is to say, about twenty times as long as a DNA molecule of molecular weight 12×10^6, which is about as large as such a molecule is likely to be. We are almost inevitably forced to the conclusion that the loop axis can simultaneously specify many copies of the protein molecule for which it "codes." Probably the simplest mechanism by which this result could be achieved is that suggested by Callan, who supposes that at each locus there may be one master gene in which the DNA is combined with something (e.g., histone or other protein) which endows it with the capacity to produce copies of itself which are sufficiently perfect

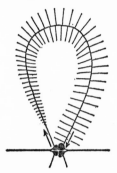

Figure 12. Diagram of loop formation in newt lampbrush chromosomes
The chromomere is represented as quadruple. On each side of the chromosome, one part of the chromomere protrudes the axial filament of the loop, and this becomes clothed with matrix as it unwinds into a loop before being wound up again into the other part of the chromomere. (From Callan and Lloyd, 1960a)

to control RNA and protein synthesis, but which cannot duplicate themselves.

A similar system of master and slave genes may perhaps occur in salivary gland chromosomes, where a single band in its puffed state may have a length of about 5 μ. Beerman (1959) attributes this length solely to the uncoiling of the gene molecule; but the magnitude is very large, and it might be that in these chromosomes a real multiplication of the gene templates occurs.

A quite different line of reasoning has led Swann (1961) to suggest the existence of master and slave genes. He points out that the proportion of DNA, RNA, and protein per unit weight is much the same in most types of animal cells; and so is the rate of growth per unit weight. But the number of genes per unit weight must be very different in different species and tissues, since it is related to cell number and not directly to tissue mass. Moreover, the amount of DNA per nucleus varies far more widely than we can suppose that the number of genes do; for instance, the haploid amount of DNA in some amphibia (e.g., Amphiuma) is about sixty times that in birds and a thousand times that in Drosophila. The only way to make sense of this is to suppose that only some of the DNA is in the form of the true genes which make up the essential genotype, while there is a further fraction of it which may be thought of as making up a set of subsidiary or copy genes.

Thus, apart from the evidence for the existence of two types of genes, structural and controlling, which we discussed in the last chapter, we find here another indication that the chromosomal apparatus may have a more complex constitution than our conventional ideas imply.

The next structure which requires attention is the nuclear envelope. During interphase the nuclei of metazoan cells are always bounded by a definite structure, although possibly this is not true of the nuclear apparatus of bacteria and of certain unicellular organisms, such as the Desmid Micrasterias, which will be discussed in Chapter 4. The nuclear envelope, when seen with the electron microscope in sections cut approximately normal to the nuclear surface, appears usually as two dense profiles between which there is a narrow paler zone. This is most easily interpreted as two relatively solid membranes separated by a fluid-filled cavity. Callan and Tomlin (1950), who gave the first detailed description of a nuclear envelope, based on the germinal vesicle of urodeles, described the inner layer as being continuous and the outer layer as pierced by a large number of pores, each about 1000Å in diameter. A porous structure has been found frequently in the other types of nuclear envelope which have been examined more recently; and it can probably be taken as one of the basic characteristics of nuclear envelopes.

There is, however, considerable variation from species to species, and even from tissue to tissue within a single species, in the detailed structure of nuclear envelopes (see, for instance, Okada and Waddington, 1959; Waddington, 1961c). Probably one should think of this universal cell organelle as having a fundamental underlying structural pattern, just as all chromosomes have the basic structure of a linear aggregate of DNA protein molecules; but in particular types of cells, the nuclear envelope may vary in the way in which this basic structure is developed to as great an extent as chromosomes do, exhibiting types which differ from one another as strikingly as do lampbrush or salivary chromosomes.

Even if we confine our attention to embryonic or differentiating cells, clear examples of this variation in structure can easily be found. For instance, in an ovariole of a Drosophila ovary, the nuclear envelopes of the nurse cells and oöcyte differ from those of the immediately neighboring follicle cells (Okada and Waddington, 1959). The nurse cell and oöcyte nuclei, which grow very rapidly (particularly the former), are bounded by an envelope which appears relatively thick and which in transverse section shows little obvious sign of a double nature. On the other hand the presence of numerous pores is easy to see; the pores appear to pierce right through the envelope, affecting both the layers which one must assume are present (Plates VIII and XI). By contrast, in the follicle cells, transverse sections show clearly a double membrane, but the occurrence of pores is by no means clear cut. In the developing retinulae of the eye, the envelope is clearly double, and its pores are much more scattered than those of the nurse cells (Plate XVIII).

Another modification of nuclear envelope structure is seen in the germinal vesicles of the polychaete Cirratulus (Waddington, Perry, and Okada, 1961). The envelope is clearly double, but the two members are separated by a comparatively wide space except in the rather numerous places in which they are apparently fused closely together. This gives the membrane the appearance of a sheet of bubbles. Moreover, the images seen in some approximately transverse section make it seem likely that there may be more than one layer of bubbles in the thickness of the sheet; in which case the envelope would have to consist of more than two membranes. A different type of development toward a thicker membrane is found in amoeba; on the inner side of the nuclear envelope there is an array of tubules about 200 μ long and set radially to the nuclear surface, packed together in a close hexagonal arrangement (Bairati and Lehmann, 1952; Pappas, 1956). A somewhat less extreme modification of the envelope is found in young amphibian embryonic cells (Plates I, II, III, IV). In these, the outer of the two membranes is lifted away from the inner, only remaining attached to it at much more widely separated places than in Cirratulus. In the derivatives of these cells, in slightly older embryos, the spaces between the two membranes become less conspicuous, and the nuclear envelope resumes a more conventional form.

The most striking modification of nuclear envelope structure tends to occur in cells, such as nurse cells and oöcytes, in which very rapid and massive synthesis of cytoplasm is proceeding, and where one might expect that the passage of information-carrying substances from the nucleus to the cytoplasm is particularly intense. This suggests that the particular form taken by the nuclear envelope has some significance in relation to this nucleo-cytoplasmic traffic. Further evidence tending to the same conclusion can be found in the relations which appear between the nuclear envelope and certain cytoplasmic structures in cells which are engaged in histogenesis. The two types of cytoplasmic organelle about which particularly suggestive evidence is available are the lamellar stacks and the membranous systems known as "ergastoplasm" or "endoplasmic reticulum."

The lamellar stacks consist of piles of membranes, each of which is composed of two members and exhibits a number of circular pores which have a thickened rim and a thinner, or even perhaps completely open, central region (Plates x and xi). These pores or annuli bear a strong resemblance to the pores seen in the nuclear envelope; and in fact, the whole lamellar stack gives the appearance, in many cases, of a pile of fragments of nuclear envelope stacked one above the other. Such structures have been seen quite commonly in certain eggs, such as those of insects and spiders, and have also been found in other types of cells. In some cases they are found quite close to the envelope of the germinal vesicle, and it has been suggested (Swift, 1956; Rebhun, 1956) that they are formed by some process of copying of parts of the nuclear envelope. In the salivary gland cells of Drosophila, Gay (1956a, b; 1959) has shown how small protrusions or blebs appear on the surface of the nucleus, and describes the formation of stacks of lamellar elements in the cytoplasm around these blebs. In other cases, however, lamellar stacks may be found at quite long distances away from the nuclear envelope. In the Drosophila oöcyte, for instance, in which they are very well-developed, I have never seen a fully formed stack in the neighborhood of the germinal vesicle; but on the other hand, one may find them at the posterior end of the egg, which is as much as half a millimeter, or even more, away from the nucleus. It seems almost certain in such cases that the lamellar stacks must increase in size without any immediate participation of the nuclear envelope, presumably by some process of the addition of new lamellar elements parallel to the already existing ones, which act as templates.

The origin of the first lamellar element remains obscure. Possibly it is always formed as a copy in the neighborhood of the nuclear envelope. In eggs, such as those of Drosophila, one has to consider the possibility that fragments of the nurse cell nuclear envelopes might form the initial centers from which the lamellar stacks grow. We know that during the degeneration of the nurse cells the nuclear envelope is particularly resistant to destruction, and may persist in the form of broken-up fragments when most of the rest of the cell has become liquefied and autolyzed (Waddington and Okada, 1960). Again, it is clear that large amounts of material from the nurse cells are passed bodily into the growing oöcyte; large holes with thickened rims can be seen in the oöcyte-nurse cell membrane at the relevant stage. There is as yet no direct evidence that fragments of the nurse cell nuclear envelopes move into the oöcyte cytoplasm in this way, but on the whole this seems rather a likely possibility and if it occurs, the fragments of nurse cell nuclear envelope might provide the templates on which new lamellar structures could be formed to build up the eventually large and many-layered stacks.

The function of the lamellar stacks is still obscure. Their membranes carry microsomal particles very similar to, if not identical with, those which lie on the more dispersed cytoplasmic membranes which are usually referred to as endoplasmic reticulum or ergastoplasm. In some cases, indeed, the membranes of the lamellar stacks extend away from the coherent structure into the general cytoplasm (Plate x), so that the stack has the appearance of being simply a more orderly region of the general cytoplasmic membrane system, rather as a grana is a more orderly region within the general structure of a chloroplast. This cytoplasmic ergastoplasm is generally held to be concerned with protein synthesis, and it may well be that this is also the main function of the lamellar stacks. It must be remembered, however, that the stacks exhibit well-marked annuli, or pores; and these are usually absent, or at least not nearly so well developed, in the more dispersed membranes of the ergastoplasm.

The relation between the nuclear envelope and the general membrane systems of the cytoplasm is perhaps even more direct than the relation between the envelope and the lamellar stacks. There is, first, a point of nomenclature to be cleared up. The electron microscope has revealed the presence of many different types of membranous structures within the cytoplasm, but there is as yet little uniformity in the way in which these are referred to. Sjöstrand (1956) refers to them all as "cyto-membranes" and then attempts to classify them into various types referred to as Alpha, Beta, Gamma, etc. Porter (1953, 1957) refers to them in general as "endoplasmic reticulum," and classifies them into the two main types of smooth reticulum and rough reticulum, the latter being characterized by the presence of microsomal particles (Palade or ribonucleoprotein granules) which are lacking in the former. The name endoplasmic reticulum was suggested when it was thought that such structures were characteristic of the endoplasm, and this can now hardly be sustained. Many authors, perhaps mainly European, make use of the word "ergastoplasm." This term was first suggested by light microscopists for regions of the cell which exhibited strong cytoplasmic basophilia, which might be interpreted as due to a high concentration of RNA. It is often used now by electron microscopists as a relatively noncommittal word to refer to the cytoplasmic structures which carry a large number of microsomal (i.e., RNA) particles. When used in this way it corresponds to the rough endoplasmic reticulum of Porter, and to the Beta cyto-membranes of Sjöstrand, but does not carry the implications which those authors have attached to their own terminology. This is the sense in which it will be used in this book.

I need scarcely remind you of the large amount of work by authors such as Porter, Palade, Siekevitz, Zamecnik, and many others, which has led us to believe that the microsomal particles are the main site of the synthesis of cytoplasmic proteins in the cell (though it is well not to forget the EM evidence which suggests that yolk may be synthesized in mitochondria and certain secretions such as milk in Golgi apparatus). The evidence is mainly of two kinds; electronmicroscopy shows that ergastoplasm provided with many microsomal particles is common in all cells known to be engaged in protein synthesis. If the cells are starved, or if the protein synthesis is halted in some other way (e.g., by hormones), the quantity of ergastoplasm is reduced, while it expands again when protein synthesis is resumed. Again, experiments with centrifugates have shown that it is the microsomal particle fraction which is most active in incorporating amino acids into proteins.

The student of development is bound to note that nearly all these experiments have been carried out with adult cells, that is to say, with cells which are already provided with a completely functional system for synthesizing a certain group of proteins. It is not clear *a priori* how far the genes are still active in such systems. Genes control the specificities of the proteins which are produced, but they might do this by bringing into being an appropriate synthetic machinery and then going out of business, leaving the machinery to be switched on or off by agents which do not control the specific nature of the end product. As I said earlier, genes are the architects of development, and there is *a priori* no reason why they need continue to function as the works managers after development has brought a factory into existence.

In our present state of comparative ignorance it is difficult to decide whether genes are continuously sending out the relevant template information for protein synthesis even in adult tissues, or whether after a certain stage of differentiation has been attained, the microsomal particles become able to operate on their own. There is considerable evidence that in cells still engaged in differentiation gene-action (presumably involving the formation of messenger RNA's and its passage to microsomal sites of synthesis) can be rather rapid, which suggests that the microsomes have little autonomy. For instance, many years ago I showed (Waddington, 1940d) that back-mutations, or some phenotypically equivalent genetic changes, of the straw locus, induced by X-rays in the cells of the pupal wing of Drosophila heterozygous for stw^3/stw^+, can cause the production of wild-type wing hairs even if they occur after the last cell division before the adult state is reached. Again, it has been known for a long time that certain characteristics of pollen grains can be controlled by genes in the short space of time after meiosis and before the maturation of the grain (e.g., the *waxy* gene in maize). Another very rapidly acting gene is the normal allele of the female-sterile *deep-orange* in Drosophila, which can have an effect within a few minutes of its entry into the egg of a *deep-orange* female (Counce, 1956). These examples—and it would be very easy to quote many others—have always seemed to me to make it likely that the cytoplasmic organelles concerned with protein synthesis in developing cells are somewhat labile in their properties. Discussing differentiation some years ago, I wrote (Waddington, 1956, p. 402): "We are already faced with the difficulty of accounting for this progressive series of changes in a system one of whose major components consists of genes which we believe to retain their identity throughout. The difficulty is only made the greater if we have to suppose that the major factors in the cytoplasm also retain their identity."

Recently, in a beautiful set of experiments, Brenner, Jacob, and Meselsen (1961) have shown that the rapidly acting genetic control of induced enzyme formation does not involve the production of new microsomal particles, but only the elaboration of new messenger RNA which changes the specific activity of already existing particles. Lamfrom (1961) has even been successful in making a similar process work in a cell-free system; microsomes prepared from rabbit reticulocytes, when incubated with an RNA fraction from sheep cells, brought about some synthesis of sheep hemoglobin, as well as rabbit hemoglobin. The appearance of the former shows that the specificity of the microsomes can be altered by soluble messenger RNA, and that of the latter, that the original specificity can persist for some time in the absence of new supplies of the corresponding messenger.

All these cases of rapid action suggest that the genes are actively sending out information-carrying materials. But they all relate to differentiating tissues. Is the situation necessarily the same in cells such as those of the adult liver or pancreas, in which the elaboration of a protein-synthesizing machinery has been completed? In experiments in which microsomes from adult or late embryonic heart muscle were inoculated on to the chorio-allantoic membrane of the chick, Ebert (1959) found that some of the cells of the membrane developed into striated elements resembling muscle cells. This would, if accepted at face value, be evidence for the retention of specificity by these microsomes; but, as Ebert points out, there is a difficulty to be cleared up. The experiment only succeeds if the microsomes are mixed with Rous sarcoma virus; and there is a possibility of some sort of "epigenetic recombination" between parts of the virus particles and RNA (presumably messenger RNA) associated with the microsomes. We need many further experiments on the intercellular transplantation of microsomes, and on the treatment of developing cells with RNA fractions, in the manner initiated by Clayton and Okada (1961).

It might be argued that it would be evolutionarily advantageous to an organism if its adult cells were protected against the disturbing influences of somatic mutation; and that this could best be done by rendering the microsomes in them independent of continuing supplies of new information from the genes. The argument would hardly apply to microorganisms, which seem able to be so prodigal of individual cells that they can rely for the evolutionary potential on such wasteful processes as mitotic recombination and the other parasexual mechanisms discussed by Pontecorvo (1959).

Another problem which suggests to me that it may be advisable to keep an open mind on this matter arises when one considers the numbers of microsomal particles in a cell, and the number of different proteins which they may be producing. The number of particles per cell can certainly be as large as a million, and probably in some cells is still larger. The number of proteins these particles are producing is probably at least a few hundred in any growing cell, and one imagines that it may in some cases be as many as a thousand. How does the information-transfer system work in such circumstances? Does any one particle continue to produce the same protein, obtaining repeated instructions from a corresponding gene; and if so, how does the messenger RNA recognize its appropriate particle? Or does a particle synthesize a protein until its RNA-messenger is used up; and then pick up any messenger, perhaps with different instructions, which happens to be around; and if so, how are the different types of synthesis kept in balance; and in particular, how are they located at the correct places within the body of the cell? (This problem will be discussed in the next chapter.) Again, what happens when a large cell is synthesizing predominantly one or a few types of protein (e.g., a muscle cell forming actin and myosin); is it plausible to suppose that the few gene loci concerned could make sufficient messenger RNA to replenish the large number of particles if the independent active life of the particles is short?

Some of these questions seem easier to answer if we postulate long-lived synthesizing particles; others if we suppose that the instructions to the particles have to be renewed at frequent intervals. We require more facts; possibly experiments with antibodies labeled in such a way as to be recognizable in the electron microscope will produce them for us. In the meantime, it certainly does not seem safe to assume that the gene-action systems which always operate in just the same manner in adult cells as in embryonic cells, and that the latter can be fully understood on the basis of studies of the former. In this, as in several other contexts, the only sure route to an understanding of differentiation is to examine the cells of embryos in which it is proceeding, technically difficult though such investigations may be.

Rather few attempts have yet been made to follow the course of differentiation with the electron microscope (a review of some of the data is by Fawcett, 1959). One reason for this may be the fact that early embryonic cells seem usually to be very difficult material to prepare satisfactorily for examination with this instrument. As is well known, they have a high water content, and in the earliest stages of development the cytoplasm is often rather bare of well-formed organelles. There are exceptions to this, for instance, in insect eggs. As we shall see later, the cortical regions of a Drosophila egg contain a large number of elaborate membrane and granular systems (including lamellar stacks), some at least of which have been derived rather directly from the cytoplasm of the nurse cells. It is not known how far the cytoplasm of other eggs, belonging to what is often called the "mosaic" type, have a similarly elaborate cytoplasm, although the few studies that have been published on molluscan and ascidian eggs (e.g., Berg and Humphreys, 1960; Reverberi and Mancuso, 1960); and our own rather superficial look at the eggs of the mollusc Limmea and the polychaete Cirratulus has certainly not revealed anything as elaborate as can be seen in the cytoplasm of Drosophila eggs. It seems rather likely, however, that eggs of the mosaic type will be at least somewhat more complex in structure than "regulation" eggs.

We have recently followed changes in ultrastructure involved in the early steps of differentiation in the amphibian embryo which is, of course, a classic example of a regulation type. In the early gastrula stage the cytoplasm is full of thin-walled vesicles and very fine granules, with rather few microsomes and little well-formed ergastoplasm (Plate I). The mitochondria are small roughly spherical in shape and with few internal cristae (Karasaki, 1959a, b). The nuclear envelope is also a simple double membrane with no very special features except that in many nuclei it extends in long finger-like processes into the depth of the nuclear mass. These long processes of cytoplasm, which are all completely lined with nuclear envelope, often contain many mitochondria. They clearly result in a very intimate spatial intermingling of cytoplasm and nucleus. If one is to suppose that they are produced by some active process, it would be difficult to avoid the conclusion that they are the signs of important physiological interactions between these two cellular components. However, the cells of the embryo at this stage are still actively engaged in mitosis, and it may be that the cytoplasmic fingers are no more than the last remnants of the spaces between the telophase chromosomes.

This suggestion raises the interesting problem of the mode of origin of the nuclear envelope during telophase. In some cells, for example, onion root tips, studied by Porter and Machado (1960b), and locust spermatocytes described by Barer, Joseph, and Meek (1960), it has been rather convincingly shown that after cell division the new nuclear envelope is formed by the coming together of ergastoplasmic membranes which have persisted in the cytoplasm at some distance from the group of chromosomes. In early newt embryos there is, however, little cytoplasmic endoplasmic reticulum, and the pictures we have seen would be more easy to interpret on the supposition that the new nuclear envelope is actually produced by the chromosomes themselves as they swell up in telophase (Plate xv). As we shall see later, there is a good deal of evidence which strongly suggests that during early differentiation the nuclear envelope during interphase gives rise to ergastoplasm. We seem, in the amphibian material, to be confronted with a process in which chromosomes give rise to nuclear envelope and that in turn to ergastoplasm; whereas, in other types of cells we have evidence of ergastoplasm giving rise to nuclear envelope which—it is not too much to suppose—may influence the activities of the chromosomes. If both these sequences of interaction actually occur, we have here a set of structural events which are, at least, an attractive candidate for the role of the carrier of the two-way information traffic which we know must go on between chromosomes and cytoplasm.

Returning to the cell as a whole, in the early stages (gastrula) there are very few of the formed cytoplasmic structures which we tend to think characteristic of cells engaged in rapid synthesis. We have followed the later stages of differentiation particularly in the notochord cells. By the late neurula stage, more cellular organelles are beginning to appear. There is first a gradually increasing development of a characteristic type of ergastoplasm (Plates III and XVI). This consists of groups of flattened

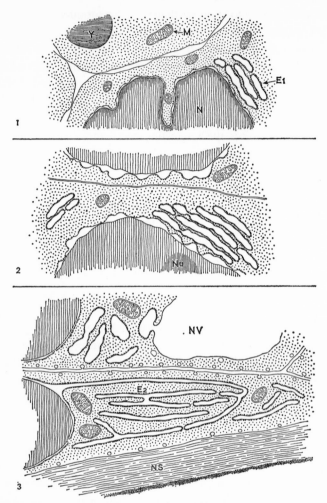

Figure 13. Diagram of main ultrastructural changes in early development of urodele notochord

In 1 are parts of three notochord cells at the early neurula stage. The nucleus (N) is deeply lobed; dense material is usually concentrated near the envelope. The cytoplasm contains few formed structures except small mitochondria (M), yolk granules (Y), and first-type ergastoplasm (E 1) located near the nuclear envelope. The cells are only loosely adherent to one another. In 2, tailbud stages, the cell membranes are in closer contact, the ergastoplasm has increased considerably, and the nucleolus (Nu) appears as a granular mass within the nucleus. In 3, one of the internal cells is above, and a peripheral cell is shown below abutting on the chordal sheath (NS) in an early tadpole stage (the internal cell is diagrammed at a some-

and apparently empty vesicles. These are lined with very thin membranes; and in the plates of cytoplasm between the vesicles, there are many small microsomal particles. Regions of ergastoplasm of this kind, which we refer to as "first-type ergastoplasm," are often found in close association with the nuclear envelope, and the appearances suggest that it is in this region that they originate. The nuclear envelope itself is by this time beginning to show some characteristic features. The outer membrane tends to lift away from the inner membrane leaving an apparently empty space (Figure 13).

As development proceeds the first-type ergastoplasm increases in volume until it comes to fill a large part of the cell. During the process the notochord cells are beginning to secrete the large internal fluid-filled vesicles which are so characteristic of them. These fluid-filled vesicles are bounded by a smooth membrane which does not bear microsomal particles. The contents of the vesicles, however, seem to be produced within the first-type ergastoplasm; and one can see images which are most easily interpreted as the vesicles of this ergastoplasm emptying their contents into the main notochordal fluid vesicles (Plate IV). In chick embryo notochords, at similar stages of differentiation, the free plasma membranes of the outermost cells, which are in contact with the general body fluids surrounding the notochord, exhibit pinocytosis vesicles at which they presumably take in fluid which is passed inward into the internal reservoirs (Jurand, unpublished). So far a similar process has not been detected in the amphibian notochord, and it seems unlikely that it can be responsible for any major part of the rapid growth of the internal fluid reservoirs to the large size which they soon attain. It is worth noting that the notochord cells at this stage contain a large number of well-developed Golgi systems, but their involvement in the formation of the fluid reservoirs remains problematical.

Fairly soon, after the fluid reservoirs have been established and grown to a fair size, the outer cells of the notochord begin to secrete the chordal sheath to which a contribution is also made by mesenchyme cells which

---

what younger stage than the sheath cell). The cell membranes are farther apart and bordered by small plasma membrane vesicles. In the internal cells, the fully formed cisternae of the first-type ergastoplasm seem to discharge their contents into the large fluid-filled vesicles (NV) which fill up much of the cell body. In the peripheral cells, the second-type ergastoplasm (E 2) is well-developed, and is continuous with the outer member of the nuclear envelope.

surround the chorda. The formation of this sheath is accompanied by the development of a very characteristic type of ergastoplasm, quite different in appearance from that which has been described above (Plates IV and V). In the cells lining the notochord itself, this second-type ergastoplasm consists of an elaborate series of flattened sacs, the contents of which is more electron dense than the general cytoplasm. The sacs are lined by strongly developed membranes which bear many well-developed microsomal particles on their outer surfaces. These particles do not seem to be scattered at random on the membranes, but are arranged in lines, spirals, or whorls. Similar arrangements have been described in some other examples of ergastoplasm or endoplasmic reticulum, for instance, by Palade (1955); and Watson (1959) has illustrated particle groups of the same general nature which, in his case, were on the surface of the nuclear envelope. The significance of these arrangements is not understood, but the fact that they occur at all is further evidence suggesting that structures of a supramacromolecule order of magnitude have to be taken into consideration in any theory of the functioning of the synthetic systems of the cell.

An important feature is that these membranes are quite continuous with the outer member of the nuclear envelope. This continuity has already been noticed by other authors (e.g., Porter, 1957, 1960; and Palade, 1956). However, they had not studied the course of differentiation which leads up to this stage; and, therefore, did not have the crucial evidence that, at the time the continuity can be seen, the ergastoplasm is increasing in size. The fact that the ergastoplasm is growing leads one directly to the hypothesis that it is being produced from the nuclear envelope. The electron-microscopical evidence for this suggestion is still, of course, derived from fixed, static preparations, but that is only to say that it is of the same kind as that on which embryology was almost entirely based until the last few decades. Careful observation of living cells with the phase-contrast microscope, in fact, reveals phenomena which should also, almost certainly, be interpreted as the shedding of ergastoplasm from the nuclear envelope. The cells, in this case, were taken from the animal pole of amphibian eggs and were not necessarily going to develop into notochord, but the processes seen in them were very similar to what would be expected from the electron-microscopical images (Elsdale and Jones, unpublished; see Plate XVI).

In the mesenchyme cells clothing the outside of the chordal sheath, a rather similar type of ergastoplasm appears at this stage. In this the cavities enclosed by the membranes are more in the form of tubes than of sacs of considerable area, and they do not fill quite such a large proportion of the cytoplasm of the cells. In other respects, however, they look in cross section very similar to the second-type ergastoplasm within the notochord. Once again, the continuity between the ergastoplasmic membranes and the outer member of the nuclear envelope is very obvious.

During these same stages several other changes are going on in the notochord cells, particularly in connection with their plasma membranes. We shall consider these in a later chapter in connection with the morphogenetic alterations in the general shape of the tissue.

Another process of differentiation to which we have paid particular attention is the development of the eye in Drosophila. During the pupal period the adult eye is gradually formed from an imaginal bud which lies near the brain of the larva. At the end of larval life, the cells of the optic epithelium in this imaginal bud are roughly onion-shaped bodies with a swollen cell body containing the nucleus and a long prolongation which runs inward to form the optic nerves (Figure 14). These cells are grouped into bundles. As far as can be detected, no cell division takes place within this system after the end of the larval period, but the cells become differentiated into at least six different types. In the adult eye each ommatidium has at the distal (exterior) end a set of four cone cells which secrete the transparent cornea. Underneath these are eight retinula cells, of which seven are quite large and form the main rhabdomeres. Each rhabdomere consists of a bundle of rather precisely hexagonal tubules (Plates xx and xxi). The proximal ends of these retinula cells are drawn out into nerve fibers. Each ommatidium is also clothed with a set of pigment cells. There are primary pigment cells toward the external surface, and these also take part in secreting the cornea. In the deeper lying parts of the eye there are secondary and basal pigment cells, which are very similar to each other in general morphology. Finally each ommatidum is associated with a set of three hair nerve groups. Each of these groups contains four cells, two of which lie internally and are connected with the interommatidial hairs on the outer side and with nerves on the inner, while the other two cells form a sheath around them. I shall discuss the detailed morphology of the

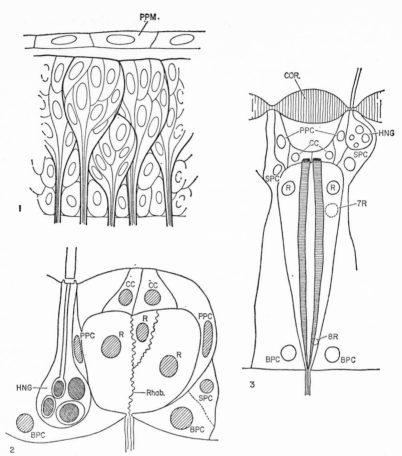

Figure 14. Development of retinulae and cone cells in Drosophila eye

In 1 is a diagrammatic section through the imaginal disc of the late larva, showing the packing of primordial ommatidial cells in the main epithelium into "bundles of onions"; at top is the thin peripodial membrane (PPM). In 2 is a longitudinal section at the 48 hour stage and in 3 at the 96 hour stage. (CC) are cone cells, (COR) cornea, (HNG) hair-nerve groups, (R) retinula nuclei (note the displaced 7th, and very small 8th in 3); (BPC), (PPC), and (SPC) are basal, primary, and secondary pigment cells. The rhabdomeres (Rhab.) begin to appear in 2, and in 3 are fully grown, provided with homogeneous tips (black). In 4 is a transverse section through a stage corresponding to that in 1, showing the complex interdigitation of the cells. Cell membranes are very prominent, even with osmic fixation, while ergastoplasm is scanty; there is some Golgi material (G). In 5 is a transverse section through the central ommatidial region at about the same stage as in 2. Rhabdomeres appear as a series of folds in the cell membranes. In the cytoplasm behind them are many small vesicles, which, it is thought, become fused with the rhabdomeres and contribute a major part of their eventual mass. The ergastoplasm is

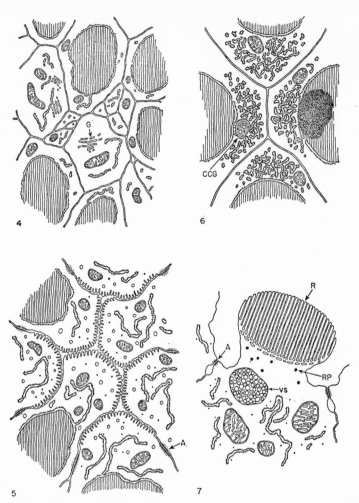

increased in volume and forms gentle undulating double profiles, clearly continuous with the outer member of the nuclear envelope. Note the attachment regions between retinulae (A). In 6 is part of the regularly arranged group of four
cone cells (at the same stage) with their characteristic ergastoplasm, which is often
concentrated near the nuclear envelope at a place where the nucleolus is also
present; each cone cell contains a cone cell granule (CCG) of characteristic type,
which lies near the nuclear envelope and sometimes seems to be connected with it
by threads. In 7 is part of one retinula at about the 96 hour stage. The rhabdomere
(R) is fully formed as a set of hexagonal tubules, on the inner border of which is
a series of small flattened boundary vesicles. Behind these are scattered granules of
rhabdomere pigment (RP). The cytoplasm contains ergastoplasm, mitochondria,
free microsomal particles, and one or a few spheroidal collections of vesicles (VS).
The attachment bodies between retinulae are near the central cavity into which the
rhabdomeres protrude, and usually have the shape shown at (A). (After Waddington and Perry, 1960)

retinular cells in more detail later on when dealing with the morphogenesis of single cells. Here I want to consider the ultrastructural modifications which are visible while the process of differentiation is going on, and for the sake of simplicity shall confine my attention to the cone cells and the retinula cells.

The cells of the larval imaginal bud, before differentiation has started, have rather scanty cytoplasm in which there are a few mitochondria but only little ergastoplasm. Just at the time when one can see the cells beginning to form their characteristic histological products (i.e., the rhabdomeres in the retinula cells and the transparent cornea lying on the outer surface of the cone cells), there is a considerable increase in the amount of ergastoplasm. In the retinula cells this ergastoplasm is very clearly continuous with the outer member of the nuclear envelope (Plate XVIII). We have in fact a situation exactly similar to that described in the amphibian notochord cells. Again, at a time when the ergastoplasm is known to be increasing in size, it is in complete continuity with the outer part of the nuclear envelope, and all the appearances suggest that it is in fact derived from it.

In the cone cells the relation between the ergastoplasm and the nuclear envelope is not so clear. It is true that the ergastoplasm seems to occur first in large quantity in the neighborhood of the envelope, but the continuity of the two membranes cannot be clearly demonstrated. In fact, in the region in which the ergastoplasm of the cone cells is close to the nuclear envelope, the nucleolus is usually present on the inner side, and immediately above the nucleolus, no clear nuclear envelope can be seen—this however may be merely a failure of adequate fixation. The situation is closely comparable to that of the first-type ergastoplasm in the amphibian notochord, in which again the ergastoplasm appears in close contact with the nuclear envelope in the neighborhood of the nucleolus.

It is, I think, important to notice that the ergastoplasms of the retinula and cone cells differ considerably in their general architecture (Plate XIX). In the former, it appears in sections as long gently undulating double profiles carrying many microsomal particles. In the latter, the ergastoplasm is made up of a tangled network of single membranes enclosing small irregular vesicles and cavities of various sizes. In the amphibian notochord there are two different types of ergastoplasms, one succeeding the other in the same cell as they secrete different differentiated products. Here we have two different types of cells, each of which produces a characteristic histological feature and each of which has its own typical kind of ergastoplasm. Such appearances strongly suggest that the morphological structure of the ergastoplasm is rather directly related to the type of synthetic process which is proceeding. But it is not clear, of course, whether the morphology is merely a resultant, or whether it has any causal influence in directing these synthetic processes. We shall return to this problem later.

There are one or two further points about these electron micrographs which are worth mentioning briefly, although they are not so closely connected with the theme of this chapter. First, the microsomes can certainly increase in number in cells which are not dividing and in which the nuclear envelope does not break down so as to release them from inside the nucleus. Their mode of origin is still obscure. They are frequently found attached to the outer side of the nuclear envelope, and possibly, they are formed there, but this is not by any means certain. A somewhat surprising observation, made during the studies on some of the mutant types of eyes in Drosophila, is that microsomal particles may be particularly concentrated in cells which are in process of degeneration; the same is probably true of Golgi material. It cannot yet be absolutely ruled out that this is a consequence of the disappearance of the rest of the cytoplasm leaving the microsomal population intact in a smaller cell mass. The appearances do not suggest this interpretation. One gets an impression that the number of microsomes is actually increasing; possibly they are producing autolytic enzymes. A high concentration of granules in degenerating cells has also been found by Lasansky and de Robertis (1961) in retinal cells of dystrophic mice.

Another point concerns the relations between characteristic phenotypic products of differentiating cells and the organelles, such as ergastoplasm, at which they are thought to be produced. In amphibian notochord cells, the first major phenotypic product is the fluid which fills the fluid reservoirs, and it looks as though this is produced directly in the interior of the cavities of the first-type ergastoplasm. In most of the other differentiation processes we have studied, however, the relation between the phenotypic products and the ergastoplasm and microsomal particles seems to be less direct. In the amphibian chordal sheath cells, for instance, the second-type ergastoplasm lies at some distance removed from the surface of the cells at which the sheath is being produced. The same is true of the Drosophila cornea, the ergastoplasm of the cone cells, and the primary pigment cells by which the cornea is secreted.

The rhabdomeres of the retinula cells are formed at the cell surface. In vertebrate eyes, the light sensitive elements are considered to be formed by the folding of the cell surface into microvilli. In the Drosophila retinula cells, the first step in the formation of the rhabdomere is also a folding of the cell plasma membrane (Plate XVII). However, it seems doubtful that the whole bulk of the rhabdomere is formed from the cell surface. Within the cytoplasm, at the time when the rhabdomere is increasing in mass, are large numbers of vesicles, which appear to be in process of joining on to the inner ends of the original folds on the plasma membrane. As the rhabdomere attains completion, these vesicles become much fewer in number; and it seems reasonable to believe that they represent the substance of the rhabdomere, which is being synthesized at an intracellular site. However, the vesicles are rarely directly connected with the ergastoplasm. Similarly, the pigment granules which form in the pigment cells of the Drosophila eye do not appear to originate in immediate contact with ergastoplasm. The same is true also of another striking phenotypic product, namely, the material forming the vitelline membrane surrounding Drosophila oöcytes. One gets the impression, therefore, that most of these striking cell constituents are not the direct products of the gene-protein systems, but are secondary elaborations rather like those we considered when discussing complementation.

Autoradiography of Developing Cells

I now want to turn to another type of work which gives us some information about the intracellular events involved in the early stages of differentiation when the genes should be active; that is, the use of radioactive labels combined with autoradiography. I shall not deal with the question of DNA duplication, but only with those aspects of autoradiographic work which are directly connected with the processes of differentiation. These have, up to the present, dealt with the synthesis of proteins and of RNA; the formation of lipids and polysaccharides, which may also be not unimportant in differentiation, has as yet been little studied.

One may perhaps begin by mentioning an odd observation on chromosomal proteins made by Sirlin and Knight (1958, 1960) in our laboratory. Drosophila larvae were fed on methionine–S^{35} for a long enough period for the proteins to be fully equilibrated with the radioactive sulphur. It was found that the total amount of residual (nonhistone) protein in each arm of the salivary chromosomes was constant from cell to cell. A fair proportion of the charge of radioactive sulphur occurred in the form of sharply bounded bands, the rest being diffusely scattered along the chromosome. The surprising fact was that the labeled bands were not the same in all the cells of a gland, but differed from one cell to the next. The observation is still not very easy to interpret. It looks, however, as though several different regions in a chromosome can synthesize, but that only a few of them do so at any one time, the synthesized material moving away from the chromosome in a fairly short period of time. If this is so, we should have to conclude that the rate of production of a protein by a particular structural gene is not directly controlled by that gene. Indeed it would appear that there must be two other controlling systems: first, one which determines which genes will be synthesizing at which time; and second, one which operates to keep the total level of synthesis by the chromosome arm as a whole to a constant level. It is interesting to find, in such physiological experiments, evidence which suggests the existence of controlling elements not entirely dissimilar from those which, as we saw in the last chapter, have been postulated on grounds of genetic analysis.

In adult protein secreting cells, amino acids are most rapidly taken into protein in the microsomal particles. This is shown by experiments on centrifugates. In early embryonic cells (at least in those of regulation types, such as the amphibia and the chick) this is not the case (Plates VI and VII). For instance, in the amphibian gastrula, after a three-hour period of labeling with phenylalanine (which is a rather specific protein label), the greatest uptake occurs in the nuclei, which at that time are only just beginning to form nucleoli (Sirlin and Elsdale, 1959). At a somewhat later stage (tail-bud), the order of protein labeling in mesoderm cells is most active in the nuclear membrane and immediately associated chromatin, then in the nucleolus and associated chromatin, then in the remaining chromatin, and finally in the cytoplasm; the situation in the chick embryo is rather similar (Sirlin and Waddington, 1956). Only at the stage of the early swimming tadpole, when the mesoderm cells are differentiating into muscle fibers, does the cytoplasm show greater activity in amino acid uptake than the nucleolus or nuclear envelope (Waddington and Sirlin, 1959). At these late stages, the synthesis of the characteristic cell proteins is underway, and one may assume that the labeling which can be found in the cytoplasm is actually occurring at the microsomes. At the earlier stages, in which the nuclear envelope and the nucleolus show the greatest activity, the proteins which are being formed are probably not those, such as myosin, etc., which are produced later in the cytoplasm. The nature of these nucleolar and nuclear-envelope proteins is not yet ascertained. In the light of the evidence mentioned above, it seems not unlikely that the activity of the nuclear envelope is connected with the formation of ergastoplasmic lamellae.

The exact role of the nucleolus is still a matter of very active debate (Sirlin, 1962). It is certainly a site at which the precursors of RNA become rapidly accumulated (Goldstein and Plant, 1955), and it seems probable that much of this RNA is actually synthesized there, rather than being merely accumulated at the nucleolus after synthesis elsewhere. Sirlin, Kato, and Jones (1961) have shown that the incorporation of nucleotides into RNA takes place first in the nucleolus-organizer or nucleolus-associated chromatin, and that the material synthesized there gradually passes into the remainder of the nucleolus (Plate vii). They found that in the early stages of development of the salivary glands in various Chironomids the nucleolus incorporates pseudo-uridine as rapidly as any of the other nucleotides; and this can probably be taken as evidence that it is forming an RNA of the transfer or soluble type; later the uptake of pseudo-uridine stops, and there is some evidence for a continuing turnover rather than synthesis of RNA in the older glands. The nucleolus is, however, almost certainly engaged in important protein synthesis, and the function of these proteins is little understood. One suggestion that has been made (e.g., Brenner, 1959) is that it is the protein of the microsomal particles. It is also possible that some of the nucleolar RNA is concerned with the duplication of chromosomal template RNA which will later pass into the microsomal particles. This RNA shows little activity in incorporating radioactive precursors and might be difficult to detect specifically in the nucleolus.

The main aspect of the autoradiographic work to which I wish to draw attention here is the evidence that there is an important change in the sites of amino acid incorporation (which presumably indicates protein synthesis) during the early phases of differentiation. The cytoplasmic sites of protein synthesis, which have been so conclusively demonstrated in adult cells, do not seem to be in operation at the beginning of development. Further, the electron microscope observations also show an almost complete absence in cells of these stages of the cytoplasmic organelles, such as ergastoplasm, which are thought to be involved in protein synthesis. These conclusions apply to eggs of the regulation type, such as those of amphibia and birds. In some mosaic type eggs (e.g., Drosophila) an elaborate cytoplasmic structural organization is present from the beginning; and we know that certain genes entering the egg with the sperm (such as, the wild-type allele of the female-sterile gene *deep-orange* studied by Counce, 1956) can produce an effect, which presumably involves the synthesis of a protein within a few minutes of their entry. Thus in these eggs the cytoplasmic synthetic machinery may already be in existence.

The control of gene-action during development has, I think, usually been considered in terms of two alternatives. The most obvious, perhaps, is that of activation, which supposes that the whole system is essentially ready to go into operation but requires some stimulus to pull the trigger. The direct alternative to this is de-repression, which implies that the whole action system is present and ready to function but is held in check by an internal agent which has to be put out of action. It is this alternative which, as we saw in the last chapter, appears to be the situation in induced enzyme synthesis. It appears also to be the mechanism involved in the less completely understood process of embryonic induction. The electron microscopical and autoradiographic evidence we have just been discussing suggests that, before either of these alternatives can come into operation, an earlier step has to be taken which can be referred to as "maturation." This implies that in the early cells the protein synthetic mechanism is not yet in readiness to be either activated or de-repressed, but requires building up before it can begin to function. As we would have said in the older embryological terminology, the cells in the very earliest stages are not yet competent. The period of competence is defined embryologically as the period during which cells will react to various agents by beginning to follow one course of synthesis or another.

Before they are competent they are not ready to react; after the period of competence is ended, they have already begun to synthesize one particular set of proteins, and it is difficult to persuade them to begin synthesizing another different set.

Although the question of how competence arises was first formulated over a quarter of a century ago (Waddington, 1934), we still know disappointingly little about it. Many competences only become manifest at late stages of development, for example, the capacity to be induced to form muscle, cartilage, cornea. Among the questions that seem most interesting to ask are the following: how far does the arising of a later competence, for example, to develop into lens or cornea, depend on the previous performance of an earlier step, such as, beginning to develop in the direction of epidermis rather than mesoderm? If an inducer for a late-appearing tissue is allowed to act at an early stage and then removed, can it perform its induction, or must it operate during the phase of competence? If it can induce, what is happening in the cells in the period between the inducer's action and the manifest response of the cells? Is the inducing substance merely stored up in the cells until something else, perhaps the microsomal population, has got ready to react to it? These questions have been asked (Waddington, 1940a) but there has been little attempt to answer them. The evidence from electron microscopy and autoradiography which we have just been discussing lends support to the idea that cells of the very early amphibian embryo have no competence at all, but require a period of maturation before they come into possession of a protein-synthesizing machinery whose specificity is then open to guidance into various alternative paths.

The Significance of Ultrastructure

The other major question which is raised by the evidence brought forward in this chapter is the significance of cytoplasmic structures at the level of magnitude of the ergastoplasm. Are they mere epiphenomena, that is causally inefficacious resultants of the basic synthetic processes? If we were confronted merely with the fact that the architecture of the ergastoplasm is characteristically different in different cells, or in different stages of the same cell which secretes a number of different phenotypic products in succession, then an interpretation of it as an epiphenomenon might be reasonably plausible. It seems to me much less so when we find the ergastoplasm apparently arising from the nuclear envelope. The nuclear envelope clearly presents some sort of a barrier which has to be penetrated by the influences (or in the fashionable modern phrase "the information") which must pass from the genes within the nucleus to the sites of protein synthesis in the cytoplasm, and also for the equally important feedback which must occur from the cytoplasm on to the genes. An actual exfoliation of the nuclear envelope to produce ergastoplasm, and the coming-together of persisting ergastoplasm to form the new nuclear envelope at telophase, as suggested by Porter, Barer, and others, seems to provide us with a most tempting vehicle to carry this two-way traffic.

Even if this were so, it would still be possible that what passes from the genes to the cytoplasm is nothing more elaborate than a number of separate molecules of messenger RNA, each capable of specifying the synthesis of one protein. The scale of magnitude of the ergastoplasm is, of course, vastly greater than that of messenger molecules of this kind, and if this is all that is involved in the genetic control of protein synthesis, then we should still be able to regard the architecture of the ergastoplasm as an epiphenomenon. Personally, I have the suspicion that this would be to oversimplify the problem. It seems to be one of the hazards of the information theory vocabulary, which is nowadays so fashionable, that it tends to make people think in terms of very simple messages, such as "Happy Birthday," as opposed to "Merry Christmas." Differentiation, I suspect, involves the transmission of messages of much greater complexity than merely "hemoglobin" or "myosin." After all, any highly differentiated cell is the result of the production of a large number of different proteins in fairly definite proportions. To specify not only the species of molecules but their relative concentrations, would seem to call for the transmission of a message more comparable to a complete paragraph. An example of the kind of structural complexity I have in mind is the way in which sequences of enzymes seem to be organized into definitely ordered groups in mitochondria. It seems rather likely that the function of many of the other subcellular organized structures is also to bring about an organized relation between individual entities whose activities it is necessary to correlate with one another (Waddington, 1961c).

Indeed, may not this principle apply to the chromosomes themselves? In the first edition of Darlington's book *The Evolution of Genetic System* (1939), he was bold enough to consider the reasons which had led to the evolutionary invention of the chromosomes. He supposed there that they had come into being because they facilitated the mechanical process of dividing the genome into two equal halves at mitosis. At the same time he pointed out that in order to reap the evolutionary benefits of recombination, it would be necessary, once a set of separate genes had been tied together into a linear chromosome, to arrange for them to become separated again by processes such as crossing over. But in point of fact, the alleged mechanical advantage is not very easy to see. There is no obvious reason why it is easier to separate the genome into two equal halves if the genes are strung together into four or five pairs of chromosomes, than when they are assembled into as many as two or three hundred pairs, or even left individually separate. In fact, in the later edition of his book (1958), Darlington seems to have had second thoughts about there being any actual mechanical advantage in the existence of chromosomes, and attributes their appearance merely to "the capacity of DNA nucleotides for unlimited polymerization to form fibers."

It seems rather unlikely, however, that chromosomes would be such a general and uniform characteristic of the hereditary substance unless they had some definite advantage over a system of separate genes. Is it not possible that the nature of this advantage is beginning to be revealed to us by investigations such as those which have uncovered the operator in the induced enzymes synthesis situation and the controlling elements in maize? The existence of structurally coherent entities on a supramacromolecular level certainly seems to offer the possibility of a control of activity of a more complex and refined kind than could easily be carried out if one had to rely entirely on soluble molecules. It seems rather likely that the possibility of such control provides the reason why such structures have been evolved both in the chromosomes and in the cytoplasmic entities, such as ergastoplasm. This, of course, implies that we shall eventually discover some type of control system which operates along the ergastoplasmic membranes in a way which is at least formally comparable to the manner in which the operator, or the controlling elements of McClintock, act on their neighbors along the chromosomes.

3. Types of
Morphogenetic Process

*I*N the two previous chapters we have seen that even if one tries to confine one's attention to the production by a differentiating system of new chemical molecular species, one finds that entities having a structural character, such as the nucleolus, the nuclear envelope, the ergastoplasm, cannot be disregarded. Indeed there is a strong suspicion that the architecture of these organelles may play an important part in the whole process. We did not, however, consider in any way how this architecture has been brought into being, and it is now time to turn to this question.

The most famous discussion of the architecture of biological systems is D'Arcy Thompson's *Growth and Form* (1942). The two concepts which he names in his title are, of course, closely allied and have, perhaps, usually been treated together. I wish, however, to try to distinguish between them, since I want in the remainder of these chapters to discuss form but not to discuss growth. The difficulty of disentangling the two concepts is twofold. The first, and perhaps minor difficulty, is the fact that many biological systems which are changing their form are simultaneously increasing in mass (i.e., growing). This is, of course, the case in the majority of embryos throughout most of their development. However, an increase in mass is clearly not an essential component to the alteration in form. In the early stages of eggs, such as those of Echinoderms or Amphibia, striking and fundamental changes in architecture occur, such as those involved in gastrulation, while it is a matter for experimental determination (and partly for definition) whether anything that can be called growth is taking place simultaneously. We can easily conceive of a mass changing in structural configuration without any accompanying alteration in total mass. The production of form does not, therefore, necessarily depend on processes of growth.

The second, and perhaps more ticklish difficulty, arises from a consideration which is in some ways the converse of that just mentioned. Granted that change of form does not necessarily involve growth: does growth always, or even only sometimes, produce a change of form? Most people would probably admit that it does not always do so, since a mere increase in size would hardly qualify to be called a change in form; but in the majority of biological systems undergoing growth, the increase in size is not the same in all directions or in all regions. In a human being, from the time of birth onward, the legs and lower part of the body grow at a greater rate than does the head, so that the proportions between the various parts of the body gradually alter. Is this to be considered a change of form? Many of the examples discussed by D'Arcy Thompson as instances of the development of form involve differential growth of this kind. Certainly very striking alterations in general appearance can be produced in this way. I should like, however, to regard these as alterations of something which I shall refer to as shape and which I shall try to distinguish from form.

It might appear, at first sight, that a distinction of the kind required could be made by saying that forms differ from one another topologically; whereas, a series of shapes could be modifications of a single topological type. However, this does not really meet the biological situation. A great many of the biological entities which we distinguish from one another by their characteristic form, such as the simple stomach of a carnivore and the complex stomach of a ruminant, are topologically equivalent, while many distinctions which are topologically important, such as the presence of a greater or smaller number of separate vesicles or other elements in a complex, may be biologically trivial. I suggested some years ago (Waddington, 1940a) the desirability of a theory of a generally topologically kind, which would be appropriate to biological forms. I suggested that such a theory would have to be in terms of "topological operators," that is notions, such as folding on a line, piercing of holes, invagination through localized regions. However, no such theory has yet been developed, and we shall have to do our best to distinguish form and shape without its aid.

I think that the clearest distinction between these two concepts that can be made at the present time is to say that a change in the form of a growing system involves the appearance of a new growth vector. The concept of a growth vector acknowledges the fact that in a biological system growth is often not uniform in all directions, but necessitates for its description not only a specification of the rate of increase, but also of the direction in which this is to take place. In the body of a quadruped, for instance, we obviously have to specify the rate of growth along the legs as well as along and across the main body. The distinction I am suggesting would imply that so long as the system possessed only these growth vectors, it would suffer only changes in shape whatever the values of the growth vectors were. The main outlines of the body and limbs of an antelope, an elephant, or a pig would differ only in shape. A change of form would be produced if a new growth vector made its appearance, for instance, in the development of a third pair of legs as in insects, or to take a less drastic example, the appearance of one or two humps on the back as in camels and dromedaries.

I am not sure that this distinction between shape and form can be made absolutely precise. One of the difficulties is that in many biological systems the growth vectors need to be described by some type of periodic function. The anterior part of the central nervous system of many early vertebrate embryos is at first a cylindrical tube which grows into a series of vesicles, which in longitudinal section have roughly the shape of a damped period function. If we were to find two brains, one of which had one more such vesicle than the other, should we consider this to be due to the appearance of a new growth vector (a change in form) or to a modification of the previous system of growth vectors which might rank only as a change in shape?

Considerably more thought is required on the general theory of biological shapes and forms. I have raised it here not with any pretentions of being able to put the whole subject into a clear-cut framework, but in the hopes of being able to clear the ground by excluding the large category of processes in which growth is obviously involved. For the purposes of these chapters, then, I shall use the word "form" to mean a structural arrangement which cannot be altered by a mere change in the existing system of growth vectors; and I shall use the word "pattern" to focus attention on the spatial interrelations between the various parts into which a form can be analyzed.

Biological forms are classifiable in various ways. In the past, the aspect of form to which most attention has been paid was its symmetry, and there have been many discussions of symmetry classes. Classifications of this kind are, however, essentially descriptive in character. If the problem of biological form is to become connected with the rest of our understanding of biology, then form needs to be dealt with from a causal point of view. There have been few attempts to consider in a systematic way the various methods by which forms can be brought into being. In the rest of this chapter I shall try to enumerate the theoretical possibilities we have at our disposal when we attempt to account in causal terms for the appearance of formed structures in living systems.

First it will be best simply to list the various categories which we shall have to discuss. They can be arranged as follows:

1. Unit-generated forms. The forms are produced by the interaction of certain unit elements, and their character is determined by the properties of the units.
   a. Particle systems: the units are treated as small volumes.
   b. Fiber systems: the units are one-dimensional.
   c. Sheet systems: the units are two-dimensional.
2. Instruction-generated forms. The structures are produced from a set of units plus a set of instructions as to how they should be assembled.
3. Template-generated forms. The structures take their basic form from some possibly simpler existing formed structure.
   a. An exact copying or simple coding exists between template and copy.
   b. Template production of noncopies, that is the coding between template and copy is highly complex.
4. Condition-generated forms. The structure arises as the working out of an initial spatial distribution of interacting conditions.
   a. Stochastic conditions.
   b. Determinate conditions.

Before discussing each of these categories and providing some examples of them, it will be advisable to mention three other dimensions of variation by which the causal antecedents of forms may be effected.

(1) The polytypic-monotypic range. Most of the formed structures dealt with by the physical sciences, such as crystals, are built of units

of only one or a few different kinds. Many of the forms met with in biology, however, arise in systems which contain many different molecular species. We tend to find that the structures we are interested in are, perhaps, largely lipoprotein, with a certain amount of nucleic acid and some carbohydrate. One of the points that it is, I think, necessary to make, is that we have at present little understanding of the nature of the form-building processes which can go on in "polytypic" systems, that is, in systems containing many different types of basic units.

(2) The synchronic-diachronic range. These two words are borrowed from the social sciences in which they are used to make a distinction between processes which operate more or less simultaneously over a short time interval, and processes which involve slow gradual alterations over fairly extended periods of time. In the present connection the distinction is between structures which exhibit their full complexity of form as soon as they appear at all, such as crystals, and others which only acquire their essential character as they progress through a series of stages to their final form. Biological forms are, as a general rule, affected by diachronic processes of development. The essential point in the present connection is whether the features which are, in a given context, being accepted as fundamental elements in the form are brought into being simultaneously or successively. For instance, consider a five-fingered hand developing out of the initially homogeneous mesenchyme of the limb bud. The essential feature of the form is the presence of five digits, rather than six or some other number. There are the two alternatives: either five condensations of mesenchyme could appear effectively simultaneously, or they appear in a definite sequence one after the other. In the former case, the form generation would be synchronic; in the latter, diachronic. And the distinction has nothing to do with the fact that, in either case, after the essential formal organization had been achieved by the appearance of five members, much further developmental modification would be necessary to bring the structure to its final adult condition. The pentadactyl pattern is probably a synchronic one, but it is clear that there are also many diachronic forms, as here defined, to be found in the biological realm.

(3) Element-elaborated and whole-controlled range. The former category includes structures which can be wholly accounted for in terms of the conditions or situations by which they are generated. In the latter

category, which includes only diachronic forms, progressive development of the form beyond a given stage is influenced by the character
which the form has already attained at that stage; that is to say, in whole-
controlled forms there is a feedback by which the present state of the
form influences its future development. We can leave out the possibility
that there might be a feedback from the final state of the form to its
actual development at an earlier stage. This is the supposition involved
in the Aristotelian concept of a "final cause," which is a teleological notion rejected by modern science. Feedback from the present state which
influences the future course of events, as that involved in an automatic
pilot, has been called teleonomic (Pittendrigh, 1958) or "quasi-finalistic"
(Waddington, 1961a), and there is no reason why such feedbacks should
not occur in any system. Feedbacks from a final cause can, of course,
also occur, but only when the final cause is an actual existing entity; as
it is, for instance, when a goal-seeking missile finds its way to the type of
target for which it has been designed. Such situations can scarcely arise
in the development of biological formed structures, but we shall find
considerable evidence of teleonomic whole-controlled processes.

With these considerations in mind we can now turn to consider some
actual examples of the various types of form.

Unit-Generated Forms

The paradigm example of a unit-generated form is a crystal of an inorganic compound. The units are atoms or ions, which are held together
in a definite order by means of valency forces. In the biological world,
we meet many examples of unit-generated forms in which the units are
considerably larger and more complex in internal structure. Indeed we
can have unit generation in which the units are cells. But before discussing the range of varieties of forms generated from units, it may be as
well to describe briefly the types of forces to which we can appeal to
explain the orderly coming-together of particles in the macromolecular
range of size, or sizes slightly larger than this.

Over and above the usual types of chemical bonding (covalent and
ionic), there are three types of forces commonly appealed to in this connection (see Waugh, 1961; Bernal, 1958): hydrogen bonding, London-
van der Waals forces, and charge patterns. A very important point is that

the forces which may be expected to arise between small atomic groupings (small molecules, or small areas on the surfaces of macromolecules)
must be very dependent on the distance of separation of the interacting
groups. For instance, hydrogen bonds attain their full strength between
atoms which are separated by about 2.4 to 3.2 Å (in an aqueous system), and their energy falls off inversely with the third power of the distance at greater separations. Again, the London-van der Waals forces
between small molecular groupings have approximately the same range
and fall off inversely as the sixth power. The result is that a macromolecule, whose surface contains a pattern of regions of specific atomic
configuration, can enter into extremely stable association with another
macromolecule which has a complementary surface which provides a
close fit; but if there is any incongruity between the surfaces, so that
some attracting regions are held a little apart from one another, the
energy of interaction will be considerably reduced. This mechanism
should, therefore, bring about very great specificity in interaction; precisely fitting complementary surfaces will behave as though they attracted
one another much more strongly than surfaces which do not fit so well.

When larger molecular groupings are in question, the London-van
der Waals forces are not so sensitive to separation, falling off only as
the inverse square of the distance. In such comparatively larger surfaces
there will also be an attraction if, as may happen in the region of the
isoionic point, the mutual polarizations between two surfaces induce
complementary patterns of charges. This attraction will fall off as the
first power of the distance, over fairly short distances, but its specificity
would be expected to be rather slight at separations which are large
enough to allow the electrostatic effects of the positive and negative
charges to interfere with each other. The forces considered here, therefore, do not provide any very convincing explanation for specific attractions operating over distances of more than a few tens of Å units, although there are some biological phenomena, such as chromosome pairing in meiotic prophase, and some examples of intercellular adhesion,
which have been thought to involve long-range specific forces. We shall
discuss these further in Chapters 4 and 5.

a. Particle systems. Crystals of globular proteins are obvious examples
of structures which are built up by the coming-together of units which
are, in effect, particles. In a protein, such as myoglobin, the tertiary

structure of individual molecules is regarded as a form generated by unit fibers (Figure 15; Perutz, *et al.,* 1900; Kendrew, *et al.,* 1960); but when these molecules are packed together into a crystal, that is a form generated by unit particles; it is monotypic, since there is only one kind of molecule present. Reconstituted tobacco mosaic virus protein sheaths without the RNA core are another rather more complicated example. When the RNA is present, we have a form in which the units are of two types, particle and fiber.

The forms produced by the coming-together of particles need not, of course, themselves have a shape at all resembling that of the units. In fact, an important instance of particle generation is the formation of fibers from roughly equidimensional protein molecules. The process has been fully studied in a number of cases. One instance is insulin (Waugh, 1957). Isolated insulin molecules have a more or less cylindrical shape, the long axis being only two and a half times as great as the diameter. Owing to specific short-range interactions between particular parts of the surfaces of different macromolecules, these can aggregate into stable complexes. The least number of insulin molecules which forms such a stable configuration is four, which become arranged as shown in Figure 15. This grouping forms a nucleus to which other insulin molecules can be added in the positions visible at the top left and right bottom corners, and later all over the surface, always becoming attached in such a way that the pattern of the basic group of four is repeated. Clearly such a process will lead to the formation of an elongated fibril, which also grows somewhat in its transverse dimension. Such fibrils may then, as units, generate other forms of the kind which will be considered in the next section.

Globular protein units, as the insulin molecules (which are actually dimers) just mentioned, can only cohere together to form large aggregates if each unit possesses two or more sites which can interact with, or adhere to, corresponding sites on other units. In insulin there are several such sites, and this leads to the possibility of the formation of elongated aggregates which may be many units thick as well as many units long. The simplest situation is one in which each unit possesses only two interacting sites. In general these will not be symmetrically placed on the surface of the globular unit. The aggregation of a number of units, each with two asymmetrically placed sites of adhesion, will produce some sort of helical structure (Crane, 1950; Pauling, 1953; see Picken, 1960). The precision with which the helix will be formed depends on the ease of rotation at the joints between the units where they adhere to one another. If the adhesion is due to a single bond, rotation will be easy and the helix rather disorderly; if the units are held together by short-range forces (London-van der Waals or complementary charges) between small regions which fit one another, rotation will be more difficult and

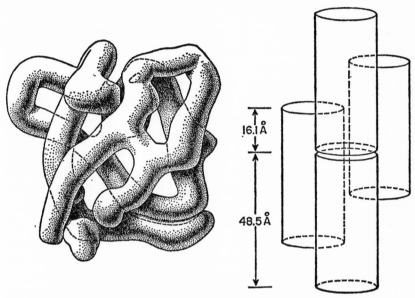

Figure 15. Protein molecules

On the left is the tertiary structure into which the fibrous myoglobin macromolecule becomes folded. (From Kendrew and others, 1960) On the right is the stable configuration taken up by four elongated insulin molecules to which further units may be added to repeat the basic pattern. (From Waugh, 1961)

the helix more perfect. In any case, the formation of helical fibers is one of the simplest performances which globular proteins may be expected to do. It probably underlies many of the phenomena which, at one time, were interpreted as the production of fibers by the unraveling of the polypeptide chains which are coiled up in the globular protein molecules. Present opinion seems to be that such an unraveling is a less likely event than cohesion of the globular units into helical polymers.

There are many more complicated structures, still on the ultramicroscopic level, which are probably also to be regarded as generated by unit particles. For instance, this type of generation seems likely to be responsible for the formation of the bronchosomes formed in the Malpighian tubules of certain insects (Smith and Littau, 1960). These have a surprisingly regular solid geometric shape, which is, however, not that of any simple crystal. The regularity is so great that one suspects that some whole-control may play a part in their formation. The EM observations have not revealed much about their genesis. They seem to make their appearance within a cloudy, roughly spherical, granule.

It is perhaps rather surprising that one does not see more evidence of particle-generated forms at the ultrastructural level. Why is it, for instance, that the nucleolus with its very high content of solid matter (mostly protein and nucleic acid) shows so little sign of crystallinity? Is it perhaps because it is too busy metabolizing; and is there some antagonism between the generation by a set of particles of a definite form and their carrying on of active metabolism?

Well ordered aggregates may be formed from particles which are considerably larger than macromolecules. For instance, Barker and Deutsch (1958) have shown that at certain stages in the life cycle of soil amoebae microsomal particles may become arranged in a crystallized order, while Goldacre (1954) has illustrated orderly arrangements of bacteria of the same general nature.

Some cell aggregates which develop a shape should perhaps be considered as particle-generated forms, the particles in this case being the cells themselves. This would be the case if the cells were held together by definite attachment bodies or desmosomes, formed at definite locations on the cell surface (Figure 43). The cells would then be joined in a way which is formally similar to the conjugation of atoms with definitely directed valencies. If, however, the cells are held together merely by general membrane adhesiveness, the form would come into the category of those produced by sheet systems.

The interactions between unit particles, by which form may be generated, is not confined to the adhering of the particles to one another. A quite different type of form generation is seen in the production of periodic structures in the well known Liesegang phenomena. Here the bands of precipitated material may indeed be crystalline—in which case their intimate structure is due to particle generation operating by adhesion between the units—but the major periodic structures which have attracted attention to the phenomena are produced by diachronic processes (involving diffusion, evaporation or the like) in a polybasic system. Nevertheless these periodic forms result from the nature of the units making up the system, and are, therefore, to be classified as unit-generated.

There are probably many other complicated types of interaction between particles which give rise to structural organization, but for the most part they still remain obscure. We may, perhaps, expect that future developments will reveal several quite new types of unit generation of form, as unlike those we now know of as Liesegang phenomena are unlike crystallization. There is not only the realm of polybasic systems to be explored, but also the many biological units, such as protein molecules, which are not static entities but possess inherent properties of movement, as contraction, spiralization, etc.; the possibility of form generation as a consequence of these properties is further discussed in Chapter 4.

b. Fiber systems. The most diagrammatic case is that of collagen, which has been thoroughly studied by F. O. Schmitt and his co-workers (Hodge, 1959a, 1960; Hodge and Schmitt, 1960; Schmitt, 1959). The basic macromolecule is a triple helix with an orderly repeating structure of amino acids, the total length of the macromolecule being about 2800 Å. These macromolecules can be brought into solution as separate entities, and from these solutions various types of fibers are spontaneously generated under various conditions (Figure 16). The fibers arise from the lateral apposition of the monomers, and also by terminal adhesion of monomers to give long lengths. There are three main types of fiber. In the native type, the ends of the monomers overlap. The amino acids which are brought side-by-side by this arrangement are not identical but are of similar type (Figure 17) .The fibers exhibit a set of cross bands (particularly visible after staining in phosphotungstic acid) which repeats at about 700 Å. In the second type of fiber, known as the "segment long spacing type," the monomers are united head to tail, with their terminal ends and all their constituent amino acids lying in register. This produces a repeat spacing of 2800 Å with many intermediate bands. In the third type, known as the "fiber long spacing," the arrangement of the monomers is essentially similar as the second type except that alternate fibers run in opposite directions. This again gives a repeat spacing

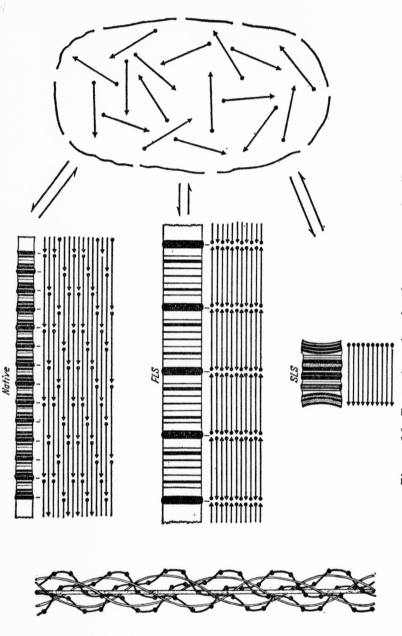

Native

FLS

SLS

Figure 16. Formation of ordered aggregates of collagen

On the left is the triple-helical structure of the elementary tropo-collagen macromolecule. On the right are ways in which molecules from solution reversibly form ordered aggregates. Below is the segment long-spacing arrangement with many molecules exactly in register; center is the fibrous long-spacing arrangement with partnermolecules head-to-tail; and above is the native type of collagen with molecules displaced in register. Observed with EM after phosphotungstic staining. (After Schmitt, passim)

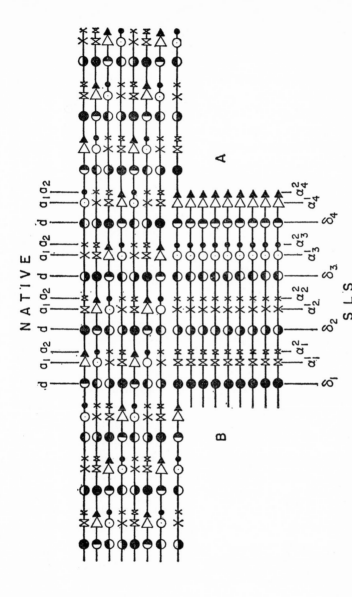

Figure 17. Diagrams of tropocollagen macromolecular packing in native type and segment long-spacing arrangement

The circles, crosses, triangles, etc. on the molecular axis indicate various amino acids. The darkness of the symbols suggests the capacities for phosphotungstic staining (only a few of the total number of amino acids are shown). (From Hodge and Schmitt, 1960)

97

of 2800 Å with few intermediate dark bands. Other elongated aggregates, in which there is no lateral alignment but the whole mass assumes a tactoid-like form, can also be produced. Finally, if the monomers are broken up with ultrasonics, certain other ordered aggregates may also be found. Some of these primary fiber-generated forms may serve as a basis for further structural elaboration. For instance, the native-type reaggregates (but none of the others) cause the precipitation at certain sites of hydroxyapatite from metastable solutions; thus, giving rise to incipient bone crystals (Glimcher, 1959, 1960). Presumably these aggregates have sites with a steric arrangement of amino acid residues which act as an effective template.

Collagen fibers are monotypic but many polytypic fiber systems also occur. The muscle fiber, which contains an orderly arrangement of myosin, actin, etc., is a good example (Hodge, 1959b, 1961; Huxley, 1961). So presumably is the chromosome with its fibers of DNA and protein. Both these structures are, of course, notably mobile, exhibiting the fast contractions of the muscle fibers and the slower, but even more extensive, coiling and uncoiling phenomena of the chromosomes. A certain dynamic quality seems often associated with fibrous systems. A dramatic example can be seen in the cytoplasm of trichocysts of certain protozoa. When at rest the cytoplasm of these cells has no visible structure in the electron microscope. On the discharge of the trichocyst, regions which in sections show periodicities develop very rapidly, in a period of a few milliseconds. The periodicity has at first a wavelength of 100 Å, then 250 Å, and finally reaches its full development at a period of 550 Å (Rouiller and Fauré-Fremiet, 1957; Grimstone, 1961).

Fiber-generated forms are probably not always simply linear. At the ultrastructural level a number of cross connected three-dimensional network arrangements have been described (Figure 18). For instance, Weiss (1956, 1959) has described, in the basement membrane underlying the skin of amphibian tadpoles, a structure consisting of alternating sheets of fibers, the fibers in neighboring sheets running at right angles to one another. In some other biological membranes, for instance, Dessermet's membrane in the eye, the fibers are definitely cross-linked into a three-dimensional network. Such structures may often be polytypic in constitution, but this is probably not necessary, since similar structures can be formed from solutions of paramyosin. Whether such forms are entirely element elaborated or involve some whole control is also not known.

A very interesting type of biological structure, built of fibers, are organs, such as flagellae and sperm tails (Afzelius, 1959; Gibbons and Grimstone, 1960; Grimstone, 1961). These have such a regular structure that it looks as though it should be easy to understand how their form is generated, but, in fact, we know very little about it. Their architecture seems always to consist of a set of nine peripheral fibers with two central

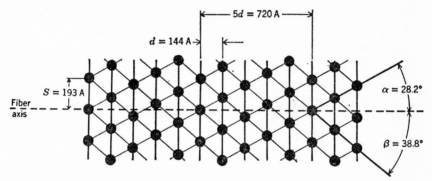

Figure 18. Cross-linked arrangement of fibers found in preparations of tropomyosin stained with phosphotungstic acid (From Hodge, 1959b)

ones (Figure 19). The nine peripheral elements are compound. Sections show them to be made up of two elongated tubular elements, one slightly larger than the other. One of these carries two longitudinal ridges which in section appear as horns. These are arranged in such a way as to make the whole structure asymmetrical. There are also certain cross connections between the eleven members of the fiber system.

It seems most probable that the system is polytypic in nature. This is strongly suggested by the fact that in certain regions (particularly the basal part and the distal tips) some elements of the pattern may drop out while others persist. This occurs in a regular way which suggests that the elements differ in composition. Flagellae always originate from a centriole or some similar structure, and it is possible that they are produced by one of the forms of template action which we shall discuss later. On the other hand, it seems perhaps more likely that they are a structure generated by unit fibers. This could be tested by seeing

whether fragments of them could reconstitute themselves into their normal arrangement, but no such experiments have yet been performed.

We have in our laboratory investigated a situation concerning sperm
tails which might, perhaps, have given a clue to their mode of formation. The water snail *Limnea peregra* is one of the comparatively few

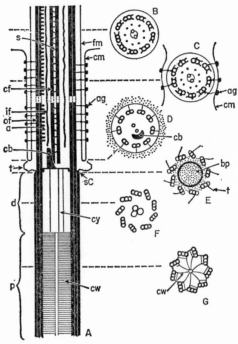

*Figure 19. Structure of flagellum and basal body in flagellate Pseudo-
trychonmypha*

In the basal body (regions p and d) are various fibrous connections among the
nine peripheral elements, which here are tripartite. The typical "9 + 2" arrangement of a flagellar axis is illustrated by section B. (From Gibbons and Grimstone,
1960).

strongly asymmetrical species in which both right- and left-handed races
are known. As was shown many years ago by Boycott, Diver, and Sturte-
vant, the asymmetry is a simple Mendelian character, controlled by a
gene which has a maternal effect and acts on the egg while it is being
formed in the maternal ovary (cf. Waddington, 1956). Individuals which
contain the factor for right-handedness, which is dominant, produce eggs,
all of which develop into right-handed individuals, whatever the nature

of the sperm by which they are fertilized; while individuals, homozygous for the left-handed factor, produce eggs developing in a left-handed way, whether or not the sperm entering them carries the right-handed or the left-handed allele. The mode of operation of the asymmetry factor on the egg is not known. One possibility, perhaps a rather remote one, is that it controls some asymmetry of important macromolecules. If this were so, it might perhaps affect the asymmetry of other cells besides the eggs, for instance, of the sperm and in particular of their tails.

In the early days of electron microscopy, Selman and I (1953) examined Limnea sperm tails, mounted either *in toto* or after ultrasonic disintegration. The most obvious structure was a system of spirals, which were seven in number on the midpieces, decreasing probably to three toward the end of the tail; there were also some fine central fibers, which we took to be ten in number. The direction of screwness of the main spirals was always the same in both right- and left-handed strains. At the time we took this simply to show that the factors controlling the asymmetry of the eggs have no effect on the symmetry of the sperm tails.

The presence of seven spiral fibers in Limnea sperm tails may at first seem a little surprising, since, following recent work on such objects with the technique of thin sectioning, we have come to associate the number nine with structures of this kind. However, there is really no contradiction here. Thin sections of Limnea have shown quite clearly that their axial group of fibers, which we had originally counted as ten, are actually the usual grouping of nine peripheral and two central ones; the former show the typical double structure with longitudinal ridges and an asymmetrical architecture of the kind which has been described in flagellae (cf. Gibbons, 1961). The apparent thick fibers we had seen in whole mounts turned out to be features of a layer of multiple sheets of material, which are wrapped round the central core of the sperm axis and are thrown into a series of screw-shaped folds; the number of these is seven near the head and decreases posteriorly. We have still not been able to prove quite conclusively that the asymmetry of the central nine plus two elements is the same in both the left-handed and right-handed races, but the available evidence tends in that direction.

These sperm provide an interesting example of the complexity, and at the same time regularity, which may be found in biological ultrastructures. How does it come about that the sheets wrapped round a nine-membered inner structure are themselves molded into a seven-membered corkscrew? The system is simple enough to provide a challenge to understanding, but it is a challenge which we are at present quite incapable of meeting. Possibly a further study of polybasic model systems, that is mixtures of substances which can as it were crystallize out in complex aggregates, will provide something of a clue.

c. Sheet systems. The formation of sheets or membranes is a very common phenomenon within living systems. The membranes themselves are probably usually polytypic, containing lipid, protein, and often carbohydrate. All these classes of compounds undergo various degrees of polymerization and enter into various forms of closer or looser combination with one another. The membranes give rise to structures which may be considered as forming two main classes: single membrane systems and multiple membrane systems.

(1) Single membrane systems. These are forms of which it is reasonable to believe that the structure has been brought into being by processes in which single membranes are the operative elements. The single membranes may, of course, be internally complex consisting of two monolayers of lipid sandwiched between two monolayers of protein, or various other arrangements of such a kind. However, in the structures to be considered under this heading the form can be considered to be brought into being by the activities of single layers, whether simple or complex, and not by the actions of large numbers of layers as will be the case in the forms to be considered in the next section.

The classic examples of single-sheet systems are the forms of certain unicellular organisms or of groups of small numbers of cells which were discussed extensively by D'Arcy Thompson (1942). He provided many examples in which the geometrical arrangements taken up by such organisms can be imitated rather exactly by the forms assumed by liquid films, such as those of soap bubbles. The soap bubble forms can in turn be accounted for in terms of surface tension. However, the question remains whether one is justified in considering that surface tension in the plasma membranes of living cells provides a fully satisfactory explanation of how they come to assume soap bubble-like forms. Most plasma membranes that have been studied exhibit some degree, though perhaps a low one, of rigidity or yield strength. For instance, when the cortical layer of a frog's egg is stressed, it wrinkles; that is to say it behaves like a solid and not like a liquid, and the notion of surface tension cannot properly be applied to it. However, this conclusion applies in connection with stresses which operate over a short period of time. If, as seems probable, the plasma membrane has some of the properties of a plastic solid, such as pitch, it might flow, behave essentially as a liquid, in response to long continued stresses however small. The application of D'Arcy Thompson's ideas to cell systems depends on the supposition that the plasma membrane has no residual rigid strength against long continued stresses; and how far this is true in practice is still a matter for debate.

Another interesting category of single-membrane forms to which D'Arcy Thompson drew attention can be produced if a drop of a protein solution is allowed to fall through a long column containing some substance which tends to coagulate the protein. In this way a surface membrane is produced around the drop, and this may "freeze" some of the complicated shapes into which the drop may be molded by the vortices and other hydrodynamic currents set up during its fall. Slowly falling drops tend to form a number of umbrella or mushroom-like shapes, some of which mimic rather precisely the contours of organisms, such as jellyfish. It is perhaps not out of the question that the forces producing these shapes in the falling drops and in the jellyfish are not too dissimilar, since a jellyfish can possibly be considered as a relatively fluid drop floating in a medium of a not very different specific gravity.

D'Arcy Thompson went on to point out the similarity in shape between medusoid or sea anemone-like forms and the transient structures which are produced in the splash which results when a drop of one fluid falls with a considerable velocity onto the surface of another. Here we meet one of the persistent difficulties of a purely descriptive approach to the analysis of form. We may find that a living system takes up a certain geometrical arrangement which we know to be characteristic of some other well understood inorganic system, but the character of the inorganic system may be such that it is quite certain that the forces responsible for the form in that case cannot be operating in the biological material. The forces which mold the body of a sea anemone cannot be the same as those which bring about the structure of a splash. We have to suppose that in the two cases we are dealing with systems which involve the same formal mathematical relations, but that the forces which are related in these ways are quite different.

In such instances we are usually still quite ignorant of the nature of the real forces in the biological case. They are probably often much more complex than would be suggested by the relatively simple mathematical description that can be given of the final shape which they bring about. For instance, to quote another example from D'Arcy Thompson, the branched antlers of certain deer lie on a surface which is curved in the three dimensions of space in a relatively simple manner. But during their development, of course, the antlers are not first formed in a flat plane and then bent into this shape; they grow by a complicated process of apposition and lengthening, and the mathematical character of the surface in which they lie must be in some way incorporated in the processes of growth which are going on at every point along the lengthening of the antler. A perhaps even more puzzling phenomenon is seen in some Foraminiferae in which a large number of highly complicated individual processes, involving the protrusion and retraction of pseudopodia, have as their result the laying down of a shell of a very simple geometrical type (Le Calvez, 1938). If the pure mathematics of such shapes gives us an idea that we are coming near to understanding their genesis, it is unfortunately being essentially deceptive.

(2) Multiple membrane systems. Biological membranes frequently show a strong tendency to arrange themselves in parallel. A simple example of the arising of such an arrangement can be seen in the nurse cells of the ovary of Drosophila homozygous for the female-sterile factor *deep-orange*. In such ovaries many of the nurse cells degenerate and autolyze. One of the most resistant elements in the cell is the nuclear envelope, which is a relatively solid-looking structure in these cells. In the degenerating cells, it becomes broken up into a number of fragments and these tend to become aligned in parallel with one another (Waddington and Okada, 1960). The formation of grana in chloroplasts is probably brought about by similar mechanisms involving the orderly parallel arrangement of already existing membranes.

In normally developing biological systems, it is rather rare to find areas of membrane ending at a free edge, such as those in the degenerating Drosophila nurse cells just described. In any membrane which is increasing in area by intussusceptive growth there must be forces of some kind or another which draw suitable new materials into the sheet and align them with the rest of the already existing membrane. At any free edge such forces are bound to be unsatisfied. There will, therefore, be a tendency for such free edges to zip together; and if this occurs, the membrane will be converted into a closed surface (Figure 20).

Closed surfaces may assume a large variety of different forms, but one can perhaps consider them as variants on three basic types; the vesicle, the disc, and the tube. The vesicle is a more or less spherical body made of a closed membrane, any two diametrically opposite points of which are too far apart to have important physical interactions with one another through the internal medium which the vesicle contains. This does not rule out the possibility that vesicles may be considerably flattened; but if the two opposite surfaces approach closely enough, they will enter into direct physical relations with one another, and at that point the vesicle becomes transformed into what is often referred to as a "double-membrane structure."

Formations of this kind are very common at the ultrastructural level. They have a large variety of general shapes. If one imagines the series of transformations just described to have started with a spherical vesicle which was able to expand in area as the two opposite sides came together, the final form produced would be circular in plan and can be referred to as a disc. However, very often we find structures built up of two membranes which lie parallel to one another over large areas, but which are both bent and folded while still remaining parallel in the three dimensions in space. In many cells the ergastoplasm has an architecture of this type.

We really require a new word to refer to this category of structures. The phrase "double-membrane," which is often used in connection with them, is ambiguous. Its most natural meaning is not to refer to the type of excessively flattened vesicle which we are now considering but rather to membranes which are composed of two closely apposed sheets of different nature. This is the case in the nuclear envelope, or on a larger scale, in such things as a lined overcoat or a piece of bread and butter. Many theoretical schemes have been put forward suggesting that most biological membranes are composed of a protein and a lipid layer in combination (Danielli, 1958). We might well wish to discuss whether

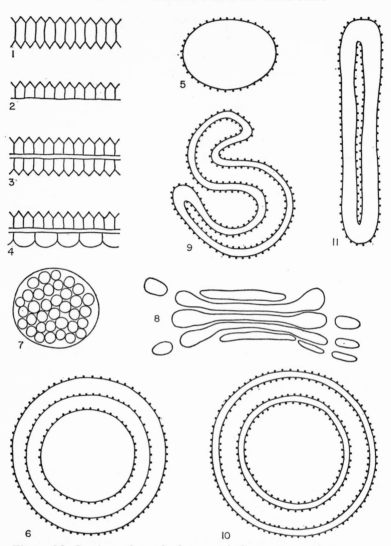

Figure 20. Sections through sheets and sheet-generated structures

1 to 4 are varieties of sheets; 1 is symmetrical with both faces the same; 2, asymmetrical; 3, symmetrical; and 4, asymmetrical, double-membranes. 5 to 8 are structures generated from simple sheets; 5 is a vesicle; 6, concentric vesicles (both from asymmetrical membranes, which may be single or double); 7, multi-vesicular body; 8, set of flattened vesicles and satellite vesicles (e.g., Golgi material). 9 to 11 are didermic structures; 9 is a folded diderm; 10, concentric diderm; 11, flattened didermic vesicle, becoming transformed into a four-layered membrane (see structure of chloroplast grana in Figure 21).

the two membranes which lie parallel to one another in the ergastoplasm are themselves truly single or are double.

Possibly the word "didermic" (two-skinned) is as simple as one can find to refer to structures in which two sheets (possibly themselves compound) are arranged parallel to one another and close enough to interact. It could also be used in a substantive form, so that we could speak of a structure being a "diderm." Clearly it would be difficult or impossible to make an absolute distinction between flattened vesicles, diderms, and complex membranes (Figure 20). When a diderm becomes extremely compressed so that the two layers lie very close to one another, we obtain what may seem to be a double membrane, but which may really be a structure composed of two skins in a mirror image relation, each skin being actually two-layered. An example is illustrated in the suggested structure of the stroma lamellae in chloroplasts in Figure 21. However, structures intermediate between flattened vesicles (or cisternae, as Palade [1955] calls them) and complex membranes are so common that it seems useful to have a name for them.

The third main category of sheet-generated structures, the tube, does not require any detailed description. It can, if one wishes, be regarded as a vesicle in which one dimension is very much elongated along a (not necessarily straight) line.

The majority of subcellular organelles are built up as complexes of vesicles, diderms, and tubes with only a few examples of free-edged membranes, such as lamellar stacks. The Golgi bodies, for instance, consist essentially of a pile of flattened vesicles. In the central regions, where the vesicles may be extremely flattened, it may be that their opposite surfaces are in such close physical association that they should be regarded as didermic structures. The peripheral parts of each vesicle, however, often look as though they were bloated by the pressure of the internal contents, and the edges of the vesicles are often ragged and appear to break down into a number of small, more or less spherical, vesicles. Mitochondria again are complex closed vesicles constructed out of a double membrane, the inner member of which may be drawn out into tubules or into what appear to be very tiny didermic structures which form the cristae. One of the simplest associations is exemplified by the multivesicular bodies, such as those illustrated from the retinulae in Drosophila eyes (Plate xx), which consist simply of a collection of small vesicle, each bounded

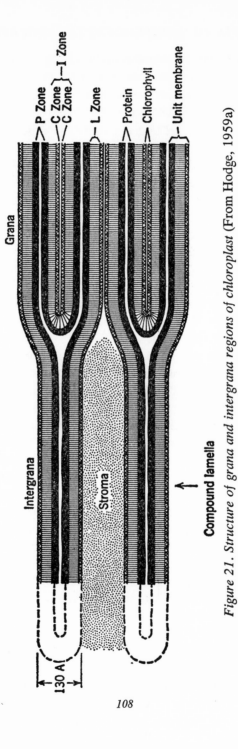

Figure 21. *Structure of grana and intergrana regions of chloroplast* (From Hodge, 1959a)

by a simple membrane, the whole collection being enclosed in another general single enveloping membrane. Another very simple arrangement of closed membranes is as a series of concentric spherical sheets like the skins of an onion. An example can be seen in the cytoplasm of a Drosophila egg illustrated in Plate IX, and they are also formed in developing chloroplasts and have been illustrated, for instance, by von Wettstein (1959). Two types of these structures can be distinguished, according to whether each of the skins is a single membrane whose inner and outer surfaces differ from one another, or whether each skin is essentially a didermic structure (Figure 20).

One of the most fully studied of the complex sheet-generated ultra-structural forms is the chloroplast. Some recent reviews are by Hodge (1959a), Sager (1958), and von Wettstein (1959). The organelle has a very similar construction throughout the whole range of plants from the simplest algae upward. There is always a limiting external envelope consisting of a double membrane. Within this there are a series of lamellae, each of which is also double, or in some places quadruple, in structure (Figure 21). In certain regions the lamellae are packed tightly together and lie precisely parallel to one another. These areas, which are usually more or less circular in plan, are known as the grana, and it is in them that the lamellae are thought to have the quadruple structure; whereas, in the intergrana or stroma areas they are only double.

The stages in the production of the fully formed structure can be studied not only by investigating the processes of development in young plants, but also by experiments involving the functioning of the chloroplasts. The lamellae seem to contain the chlorophyl as well as protein, phospholipid, and carotenoid pigments. In etiolated plants (grown from seed in total darkness) no chlorophyl is produced, and the chloroplasts, instead of containing well-developed lamellar structures, are filled with a mass of small more or less spherical vesicles. On putting such a plant into the light, chlorophyl is synthesized as can be seen by the gradual greening of the leaves; and as this occurs the vesicles within the chloroplasts appear to fuse together to form sheets which arrange themselves into the lamellar stacks characteristic of the mature chloroplast.

Wettstein has given a diagram (reproduced in slightly modified form in Figure 22) which shows somewhat similar changes going on in ontogenetic development. In plants such as barley, tomato, aspedistra, the

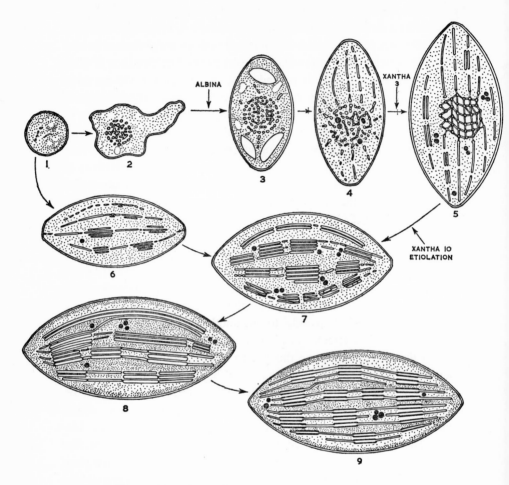

Figure 22. Chloroplast development in higher plants

The proplastid (1) can, grown in the light, develop directly into a stage with parallel flattened vesicles (6). In the dark, a plastid center is first formed (2 and 3), which may develop into a regular honeycomb network (crystalline center, 5); elongated flattened vesicles gradually appear (4 and 5), and eventually converge with the direct development on a stage indicated in (7). The fully formed chloroplast is derived by increase in the number of vesicles, their arrangement into stacks in the grana, and closer apposition of their walls. The dark spots are globuli, which contain chlorophyll and/or carotenoids pigments. The sequence of development is arrested in the barley mutants *albina, xantha 3,* and *xantha 10* at the steps shown, and in etiolated plants grown in darkness it also ceases at the point indicated. (After von Wettstein, 1959) chloroplast takes it origin from a proplastid, which at first is a small body bounded by a double membrane and containing a few vesicles. This may develop in either of two ways. The original vesicles may increase in area and line themselves up into double-membered lamellae; they then develop into the fully formed chloroplast in a rather direct manner. On the other hand, the process of organization may proceed more slowly than that of increase in membranous material. In that case, a plastid center is formed consisting of a large accumulation of small vesicles. The more peripheral vesicles gradually fuse to form lamellae, while the central area may develop a very interesting structure which appears as a net-work in thin section and which is referred to by Wettstein as the "crystalline center"—a name which is not perhaps very appropriate since it seems to consist of a mass of tubules which become arranged in a rather regular honeycomb-like structure. Etiolation, or growth at low light intensities, tends to produce this second type of development involving the formation of a plastid center.

Wettstein describes the basic processes in the development of the chloroplast as involving six stages: (1) the synthesis of the vesicles; (2) their fusion to form flattened discs; (3) the arrangement of these discs into parallel layers; (4) the multiplication of the lamellae so produced; (5) the growth and fusion of the lamellar discs to form a continuous lamellar system; (6) the differentiation of this into grana and stroma regions. The system in which these processes go on is certainly a polytypic one in the sense that it contains several different substances whose interactions are important in causing the morphogenetic changes. Undoubtedly, one of the most important substances is chlorophyl itself. Not only does the morphogenesis fail if chlorophyl is absent because the plant has been grown in darkness, but the development of the chloroplasts also comes to a halt at an early stage in plants which show the effects of mutant genes which reduce chlorophyl synthesis. For instance, in Chlamydomonas, Sager (1958) has described a yellow mutant in which chlorophyl is almost completely absent, while the content of carotenoid pigment is not reduced; in these mutant cells the chloroplasts contain only a mass of vesicles with no lamellae. In a pale green mutant of this form, the chlorophyll content is reduced to about 5 per cent of its normal value with the carotenoids almost completely absent; and in these cells the chloroplasts contain a reduced amount of lamellar material, which seems to be roughly in proportion to the amount of chlorophyl present, the level of carotenoids being of little importance.

Similar reductions in the development of chloroplasts have been described in higher plants containing chlorophyl mutants. In the diagram of Figure 22, von Wettstein has indicated the steps of morphogenesis which are inhibited by three such mutants in barley: *albina, xantha 3,* and *xantha 10.* Investigation of the development in *xantha 3,* however, shows that other substances than chlorophyl itself are involved. In this mutant both chlorophyl and carotenoids are synthesized in the very early stages of development; nevertheless, the formation of the lamellae in the chloroplasts fails, and it is subsequent to this failure that synthesis of chlorophyl ceases. Presumably therefore, the primary effect of the mutant is on some substance other than chlorophyl, which is important for the generation of the complete chloroplast structure; and it also seems to follow that the continued synthesis of chlorophyl depends on the formation of a normal lamellar system. This is a good example of a type of feedback relation which one suspects is rather common in biological materials; the presence of chlorophyl is necessary for the development of a particular structure, and the formation of that structure is necessary for the continued production of chlorophyl.

Another example of form generation by a membrane system is provided by the development of the rhabdomeres in the eyes of insects such as Drosophila (Waddington and Perry, 1960; Plates XVII and XX; and Figure 14). The structure produced is a more complex one than mere piles of flat lamellae or discs. It consists of a very regularly arranged collection of small tubular elements, which are packed tightly together into a honeycomb structure, which is perhaps like the crystalline centers described by von Wettstein in chloroplasts, although the rhabdomeres are considerably larger in scale and more perfectly developed (cf. Fernandez-Moran, 1958). The first sign of their appearance is an undulation of certain portions of the plasma membranes of the retinula cells. This gives rise to the formation of a number of very irregularly shaped microvilli. During the third day of pupal life the amount of rhabdomere material rapidly increases. It is possible that the whole of this increase is brought about by growth of the plasma membrane, giving rise to larger and deeper microvilli. However, in the cytoplasm lying below the microvilli, between them and the nucleus, there are very many small vesicles;

THE PLATES

PLATE I.[1] Most of the area is occupied by parts of one cell from newt *Triturus alpestris,* early neurula. Two other cells appear at the top right and bottom left corners. The nucleus of the cell is highly lobulated so that it appears in the section in two separate parts. Over most of the nuclear surface, the two components of the envelope are closely apposed to one another. The cytoplasm contains large oval yolk granules and irregularly shaped lipid drops (both are somewhat creased in the sectioning). There is also a fair number of small mitochondria. The ground substance of the cytoplasm fixes in the form of many small vesicles (seen toward the bottom left) but there is little or no ergastoplasm (see Karasaki, 1959a, b). (x 6,000)

PLATE II. Three stages in the development of the intercellular membranes in the Urodele notochord. Top photo is part of two cells at stage 18 in *Triturus alpestris.* A considerable gap is between the cells; and the cytoplasm shows little structure except for very small granules, oval yolk platelets, and irregularly shaped lipid granules. (x 10,000)

In the middle photo is a section from rather later Triturus cells at stage 22. The cell membranes are closely adherent. The lower part of the section is occupied by the nucleus; the outer member of the nuclear envelope is lifted away from the inner member, forming a series of cavities or vesicles. From these the first-type ergastoplasm will be derived (see Plate III). (x 30,000)

The bottom photo is a section passing approximately normally through the boundary between two cells in the inner region of the notochord of Pleurodeles at stage 36 (young tadpole). The cell membranes have separated again from each other and are accompanied by numerous cell membrane vesicles. There

---

[1]All the electron micrographs in these plates were made from preparations prepared by my assistant, Miss Margaret M. Perry, for whose care and cooperation I should like to express my thanks. They were taken either on a Siemens Elmiskop 1 or a Philips EM 75. The magnifications quoted are, in most cases, approximate only.

seems to be too many of these for them to be engaged in pinocytosis, since when they first appear, just after the stage in the middle photo, there is little fluid between the cells which could be engulfed in this way. It therefore seems likely that they are excreting fluid and that this is the reason for the gap between the cell membranes. Note the cytoplasm of the cells now contains many very fine fibrils, possibly of a pre-collagen. At the top of the photo is part of a mitochondrion. (x 57,000, stained uranyl acetate)

PLATE III. First-type ergastoplasm in early notochord cells of Triturus. The upper photo shows small amounts of ergastoplasm forming in the neighborhood of the nucleus, which lies to the lower left. In the center is some Golgi material, and five mitochondria are visible. Stage 22. (x 36,000, stained lead hydroxide)

The lower photo shows a more considerable development of the ergastoplasm. Note the arrangement of the microsomal particles in whorls and spirals at the left side where the section is more nearly tangential to the lamellae. Two pores in the nuclear envelope are clearly seen below and just to the left of the two mitochondria at the right. Stage 22. (x 48,000, stained lead hydroxide)

PLATE IV. The top photo is a cell from a notochord of Pleurodeles at stage 23. The cisternae of the first-type ergastoplasm protrude at the points marked (X) into the large fluid-filled, smooth-walled central vacuole of the cell (FFV), and it seems probable that they break down releasing their contents into it. Note the large vesicle between the two members of the nuclear envelope at (E). (x 20,400, stained uranyl acetate)

In the center photo, part of the chordal sheath (inner layer) is at the top left (S). Most of the picture is taken up by a notochord cell lying against the inner side of the sheath; parts of the membranes bounding the cell are seen above and below at (ICB), accompanied as usual with small cell membrane vesicles. The cell contains well-developed second-type ergastoplasm with the cisternae filled with material which is more electron-dense than the general cytoplasm. Two mitochondria lie outside the cisternae. At the lower part of the picture is part of one of the inner cells of the notochord. This contains a large, fluid-filled vesicle, only a small portion of which is seen (FFV). The walls of this vesicle are smooth, that is, do not carry microsomal particles. Some vesicles (V), larger than the cell membrane vesicles, appear to be discharging into the central vesicle; they are presumably remnants of the first-type ergastoplasm. Young *T. alpestris* tadpole. (x 20,400, stained uranyl acetate)

The bottom photo is a mesenchyme cell lying against the outer side of the chordal sheath (which can be seen at the bottom) in an embryo similar to that in the central photo. The ergastoplasm is in the form either of tubules or flattened vesicles with a fairly large number of holes through them. Note the continuity of the ergastoplasm membrane with the outer member of the nuclear envelope. (x 25,500, unstained)

PLATE V. Second-type ergastoplasm in a cell outside the chordal sheath in Pleurodeles at stage 36. In the central part of the photo, where the section is nearly tangential to the ergastoplasmic membranes, can be seen the arrangement of the microsomal particles in whorls and lines. (x 60,000, stained lead hydroxide)

PLATE VI. Top photo shows the incorporation of methionine-S^{35} predominately into the proteins of the nuclei in an early gastrula of Triturus. The blastopore is to the upper right, and the incorporation is most intense into the nuclei of the invaginated mesoderm and overlying ectoderm in this region. The lower photo shows the incorporation of phenylalanine-C^{14} into somite mesoderm of a Triturus embryo at the late tailbud stage. At this time the muscle fibers are beginning to differentiate, and the cytoplasmic proteins are as much, if not more, labeled than those of the nucleus. For labeling experiments which are more precisely comparable with each other, see Waddington and Sirlin, 1959.

PLATE VII. The nucleolar uptake of protein labels in early embryonic mesoderm cells is shown in the upper row. The two photos on the left show a nucleus of a mesoderm cell of the toad Bufo at the late neural stage after exposure to methioline-S^{35} for 70 mins.; that on the left is focussed on the tissue and shows the nucleolus (arrow), while the center photo is the same field, focussed on the emulsion, showing grains concentrated over the nucleolus. The photo to the right shows another cell from similar tissue after an exposure of 4 hrs. 40 mins. to labeled methionine; there is a clear concentration of grains over the nucleus, and particularly over the nucleolus (which is somewhat out of focus, below the emulsion layer). Courtesy of Dr. T. Elsdale

The two lower photos show incorporation of guanosine-3H into the nucleolus of salivary gland cells in the Chironomid Smittia. In this organism the chromosome passes right through the center of the nucleolus which surrounds it on all sides. The left-hand photo shows the incorporation after 30 mins. exposure to solution containing the tracer; the latter becomes incorporated into the nucleolus-associated chromatin. The right-hand photo shows the result of a 30-min. exposure to tracer followed by culture for 120 mins. in solution containing nonradioactive guanosine; the RNA containing the tracer has moved from the central location on the chromosome to the periphery of the nucleus. Courtesy of Dr. J. L. Sirlin

PLATE VIII. Nuclei of the nurse cells in the Drosophila ovary. The upper photo is at stage 8, when the major phase of growth is just beginning. There are already many masses of nucleolar material and accumulations of electron-dense substances just outside the nuclear envelope. (x 6,000, fixed osmic, stained permanganate)

The lower photo shows part of a nurse cell at the height of the growth phase (stage 11). The complexly folded nucleus takes up much of the central part of the picture. It is bounded by a greatly expanded nuclear envelope, which is studded with many pores, and is filled with loose masses of electron-dense material, which is made up of both DNA and RNA and represents unaggregated polytene chromosomes and nucleolar substance. The cytoplasm contains mitochondria (oval in section) lipid granules (stellate), and ergastoplasmic elements, which seem to be tubular in structure. (x 15,000) (Part of the lower photo was reproduced in Okada and Waddington, 1959.)

PLATE IX. Cytoplasmic structures in the posterior region of an almost full-grown oöcyte of *D. willistoni*. At the top left is part of the vitelline membrane. Below this the cortex contains elaborate folds of the external plasma membrane. In deeper-lying regions of the cytoplasm are various membranous structures, including a set of concentric sheets. The holes in the section are extracted lipid granules. (x 20,000, stained uranyl acetate)

PLATE X. A stack of annulated lamellae in the cytoplasm of a late oöcyte of *D. willistoni*. Some of the lamellae extend out into ergastoplasm-like structures. The dark oval profiles are mitochondria. (x 27,000, stained uranyl acetate)

PLATE XI. Pores in membranous organelles of the *D. willistoni* oöcyte. Above left is part of a nurse cell nuclear envelope; nuclear interior is to lower right, cytoplasm to top left. Note the frequent presence of a dark granule in the center of each pore and the electron-dense rim round the pore (see Watson, 1959).

The photo to the lower right is approximately tangential to the lamellae of a stack similar to that in Plate X. The pores are very similar to those of the nuclear envelope but contain no central particles. (both x 52,800, and stained uranyl acetate)

PLATE XII. Myelin figures produced during the expansion of a grain of lecithin in contact with water. The upper photo shows the lecithin-water edge at an early stage with many "sprouts," that is, thin tubules which become rolled up into a roughly globular knot at their outer tips. Some zigzags are seen (rather out of focus) at the lower right. (x 150)

The two lower photos show more central regions at a middle stage. They illustrate the rather regular and orderly structures which can be formed even in such a simple system. (x 300) Courtesy of Dr. G. G. Selman

PLATE XIII. Myelin figures. The upper photo (dark-ground phase illumination) shows the lecithin-water edge at a middle stage. Most of the tubules now run tangentially rather than radially. Note the pattern in the central area.

The lower photos are at late stages. On the left, the water edge again shows radially-directed forms with thin stalks and swollen tips. On the right is a central region of the late stage, exhibiting a resemblance to a cross-section of a myelinated nerve trunk. (all x 300) Courtesy of Dr. G. G. Selman

PLATE XIV. The top left photo is a section passing just under the surface of the nucleus lying in the isthmus of an adult nondividing Micrasterias; the folds of the isthmus are seen at (I); part of the chloroplast is at the top left. The surface of the nucleus is folded, and some of these folds are cut nearly tangentially. Note that a complete envelope surrounds the whole periphery of the nucleus. Within the nucleus are many scattered dense bodies, only a few of which are seen in this section; these are presumably nucleolar masses. (x 4,500, stained lead hydroxide)

To the right is part of the nuclear envelope of the same individual as in the previous picture, shown at a higher magnification. There is a well-developed structure of pores or annuli, many of which have a dark granule at the center. (x 60,000, lead hydroxide)

The lower photo is a section through the two new half-cells in a dividing Micrasterias at stage 4 in Figure 35. The two dark masses at the two sides are the nuclei, which seem to be extending out in thin strands which merge into the dark strands in the cytoplasm. There is no obvious organization of the cytoplasm into three main axes, and the new cell walls also do not reveal the presence of three favored directions of expansion. (x 3,000, fixed osmium tetroxide, stained uranyl acetate)

PLATE XV. The upper photo is part of the nucleus of a stage 4 dividing Micrasterias at higher magnification. Within the dark finely granular nuclear material at the upper right, little structure can be detected (except for the presence of nucleolar masses, which are not seen on this section). The nuclear envelope seems to be just forming again after the division; it is well developed in the concavity near the middle of the right-hand margin (NE), but in the processes pushing out from the nucleus toward the lower left, little sign of any definite envelope can be seen. The large dark granules with circular outlines, which are common along the nuclear projections, are particularly emphasized by lead staining. (x 32,000, lead hydroxide)

The lower photo is part of a telophase nucleus in an epidermal cell in the tail of a Pleurodeles (urodele) tadpole (part of another cell fills the bottom left-hand corner, the intercellular boundary being at [ICB]). The chromosomal materials appear as dark masses. Closely attached to several of these are pieces of double membrane (M). There are few free membranous elements not attached to the chromosomes, although there is a little Golgi material at (G).

Note the mitochondria lying among the chromosomes. (x 33,000, stained lead acetate)

PLATE XVI. The top photos, A, B, C, and D, taken within a total period of about a minute, show the formation of transient vesicles (arrows) in the neighborhood of the nucleus (N) in a cell from the caudal end of the neural tube of a neurula-stage *Xenopus laevis* embryo. It seems probable that these vesicles are the structures which appear in electron-micrographs as first-type ergastoplasm (see Plate III), but the correlation between the *in vivo* and electron microscopic appearances has not yet been definitely established. (x 1,500, from a microculture of disaggregated cells; medium Steinberg's solution plus 0.1% bovine plasma albumen pH 7.8; phase contrast) Courtesy of Dr. K. Jones

At the bottom left is a lower-power EM photo of the upper surface of the neural groove of Triturus. Note the layer of electron-dense material lying just below the surface. This is well developed just at the time at which the groove is becoming folded-up, and may play a part in the morphogenetic movement. (x 8,000) Courtesy of Dr. G. G. Selman

To the right is a membrane knot (K) in an early cleavage stage of the egg of *Limnea peregra*. The knot lies at a place where three of the loosely aggregated cells come together and consists of a close folding-together of the cell membranes. (x 7,800, from Waddington, Perry, and Okada, 1961a)

PLATE XVII. The central region of an ommatidium in a 56 hour wild-type *D. melanogaster* pupa. The margins of the seven retinula cells are deeply infolded in the regions in which the rhabdomeres will form. In the cytoplasm behind the folded regions are many small vesicles (V) which, it is thought, become attached to the folded membrane and contribute a large part of the mass of the final rhabdomere. (M) are mitochondria. Note the attachment regions of the cell membranes just outside the folded parts (e.g., at [X]). The section is tangential, that is, parallel to the outer surface of the eye. (x 39,000, fixed in OsO_4, embedded araldite, stained lead hydroxide)

PLATE XVIII. At the top is a tangential section through part of a group of retinula cells (to the left) and the neighboring secondary pigment cells (to the right) in a 56 hour pupa. In the retinula, note the shape of the ergastoplasmic lamellae. The section is almost tangential to the nuclear envelope, and it can be seen that this is provided with scattered pores with well-developed annular edges (P). In the pigment cell the large dark angular bodies are lipid, the small dark round bodies are early pigment granules. (x 19,800 osmic-araldite-lead)

At the bottom is part of an ommatidium of the same age, fixed in permanganate. This fixation emphasizes the nuclear envelope (NE), ergastoplasmic membranes and the small vesicles (V) in the cytoplasm; the rhabdomeres (RH) themselves often fix badly, while the nuclear contents show no structure and microsomal particles cannot be seen in the cytoplasm. Note the continuity of ergastoplasm with nuclear envelope at the places marked (X). At (PC) is part of a pigment cell; (M) are mitochondria. (x 36,000, permanganate and araldite)

PLATE XIX. At the top are two examples of the continuity between the nuclear envelope (NE) and the ergastoplasm (E), in 56 hour retinula cells. That on the left is fixed is osmic and shows the microsomal particles, that on the right, in permanganate. (both x 60,000)

At the bottom is a section through the central region of the four cone cells in a 56 hour eye, fixed in permanganate. Note the regular arrangement of the cells, and the character of the ergastoplasm, which differs markedly from that of the retinula cells seen in the lower part of Plate XVIII. (x 36,000)

PLATE XX. The upper photo is a 90 hour retinula, fixed in permanganate, showing the cytoplasmic vesicles uniting with the inner side of the rhabdomere. (x 60,000)

The lower photo is an adult retinula. The rhabdomere is bounded on its inner side by small rhabdomere boundary vesicles. Beneath these there is a region of sparse cytoplasm with little electron-dense material. In this region small granules of pigment (rhabdomere pigment granules) are found; four of them are visible here. There is a large retinula vesicular spheroid left of the center, and further to the left is some of the ergastoplasm with plentiful microsomal particles. (x 60,000, fixed osmic)

PLATE XXI. Low-power view of a group of eight ommatidia in a 90 hour eye. Fixation was in permanganate. Ergastoplasmic membranes are prominent in the retinula cells, but very scanty in the pigment cells. The lipid inclusions in the pigment cells are badly fixed, appearing as dark-walled holes. Note the regularly repeated pattern of the rhabdomeres. Less obvious, but in fact just as well developed, is the regularity of arrangement of the pigment cells. For a somewhat similar section of slightly older osmic-fixed material, see Waddington and Perry, 1960; Plate 3; Figure 14. (x 10,000)

PLATE XXII. The upper photo is a low-power view of some ommatidia in a 96 hour *rough* eye. Note both the ommatidia in the upper left have eight rhabdomeres instead of the usual seven; but in spite of this abnormal number, the pattern of their arrangement is repeated rather precisely. Other ommatidia in the same eye have greater abnormalities; for instance, one with only three rhabdomeres is in the bottom left corner. The fixation (by osmic) is not adequate for fine cytological detail. (x 5,580)

In the center photo is another ommatidium from a 96 hour *rough* pupa. Five retinulae make up an ommatidial axis (with two retinulae protruding into the center instead of only one). To the left, another three retinula cells have each developed some rhabdomere material. In cell 1 this extends round a great part of the whole periphery, while in cells 2 and 3 it is also very extensive. The rhabdomere material on the right side of cell 1 abuts directly against a pigment cell; there is no space (fluid-filled or otherwise) above it. Some of the rhabdomere material to the left of cell 1 is fused with that of cell 3. (x 7,440)

At the bottom is a low-powered view of a transverse section through some ommatidia in a 96 hour *Glued* eye. The anatomical arrangements are so disturbed that it is almost impossible to identify separate ommatidia. The bodies of rhabdomere material are very massive, and in most, but not all cases, have an empty-looking space lying above them, corresponding to the interrhabdomere space in a normal ommatidium. (x 5,115)

PLATE XXIII. The upper photo shows rhabdomere material in a highly disarranged *Glued* eye. These cells actually lie below the basement membrane of the eye. (x 20,000)

The lower photo shows part of a 96 hour ommatidium from a *polished* pupa. The retinula cell taking up the center part of the photo is necrotic. The structure of the rhabdomere material has been largely lost, but the cell contains an unusually dense concentration of microsomal particles and mitochondria. Another degenerating cell, probably from another ommatidium, is at the bottom right, separated from the first by two healthy pigment cells. It also contains a large number of microsomal particles. (x 24,000, fixation in osmic)

PLATE XXIV. At upper left is a cross-section of the rhabdomere tubules in a 96 hour *Glued* eye. The size of individual tubules and the ordering into rows are both considerably more variable than in a normal eye. (x 60,000) At the top right is a single retinula cell forming a moderately normal rhabdomere with space above in a *Glued* eye. (x 24,000) In the center is a much more extreme disarrangement of material in one of the rhabdomeres of a *polished* eye. (x about 22,000)

The lower photo shows a 96 hour *polished* eye fixed in permanganate. There is an irregularly-shaped retinula multivesicular spheroid at (MV), while at (R), two retinula cells buried amongst the pigment cells have produced small amounts of rhabdomere material together with typical retinula pigment. (x 24,000)

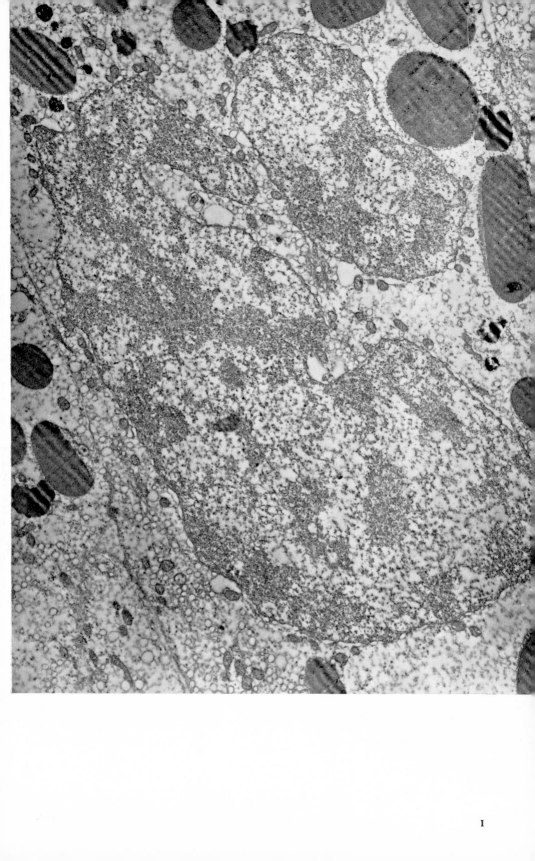

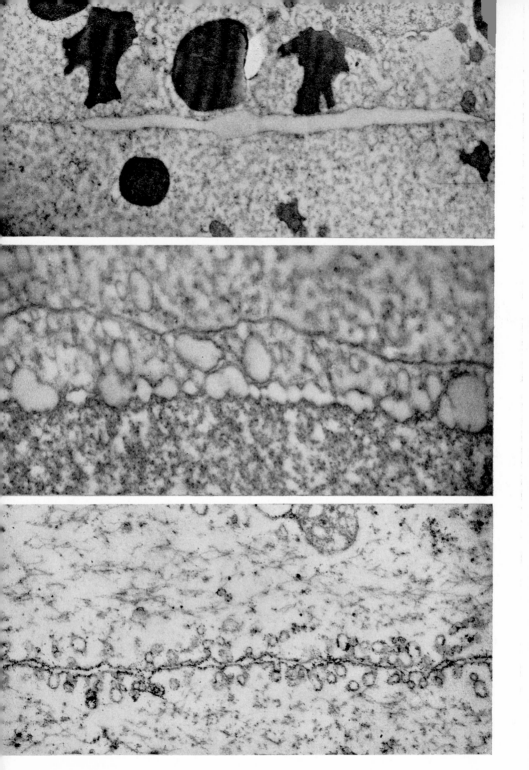

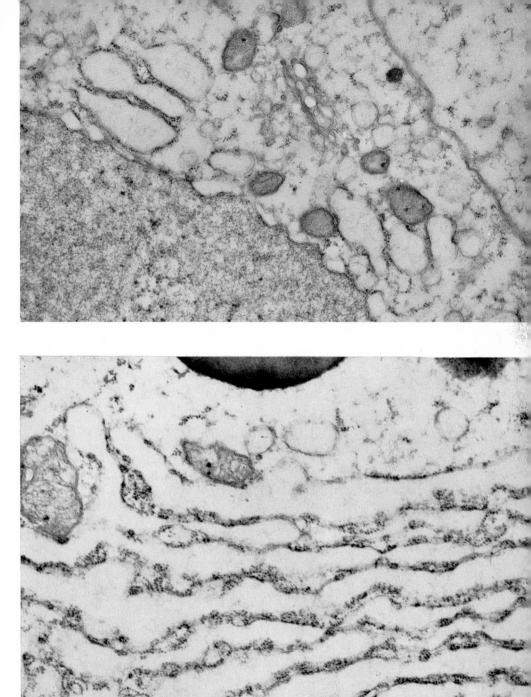

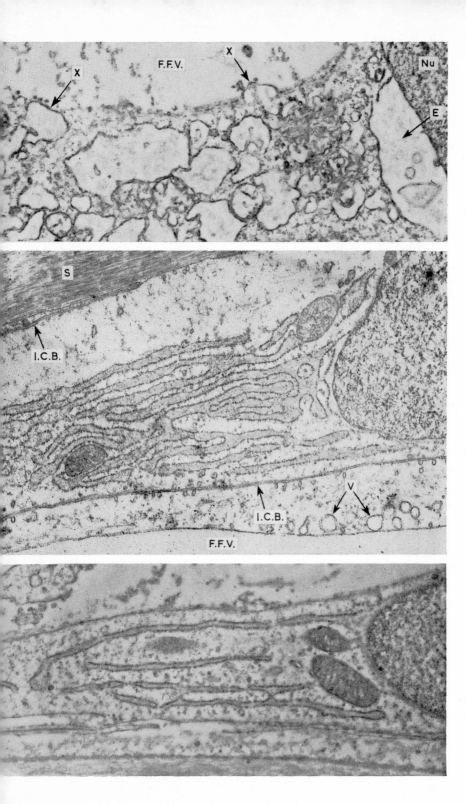

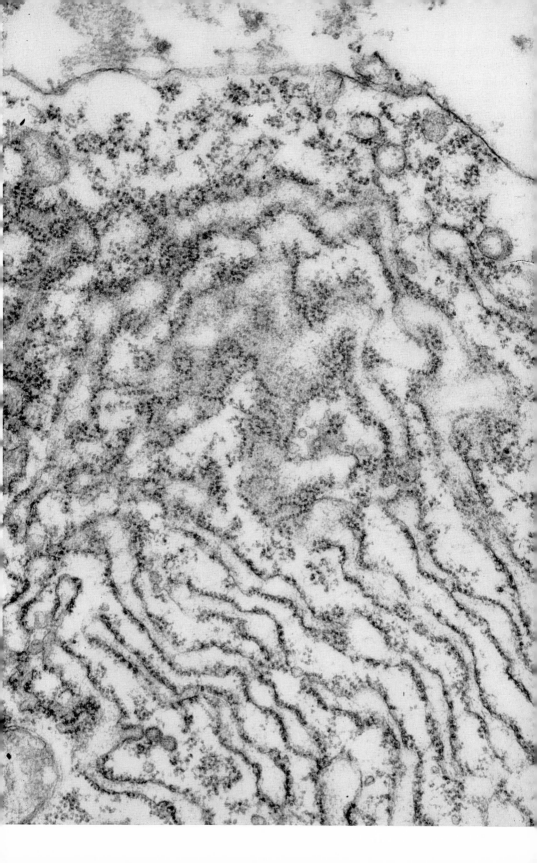

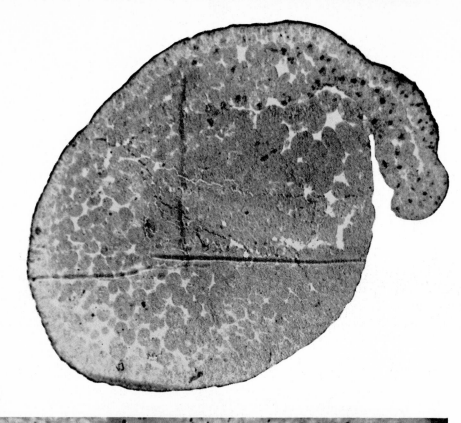

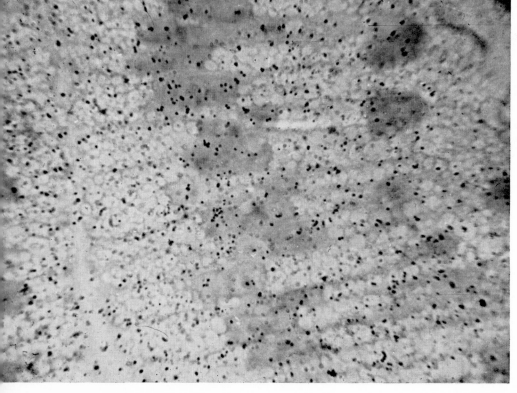

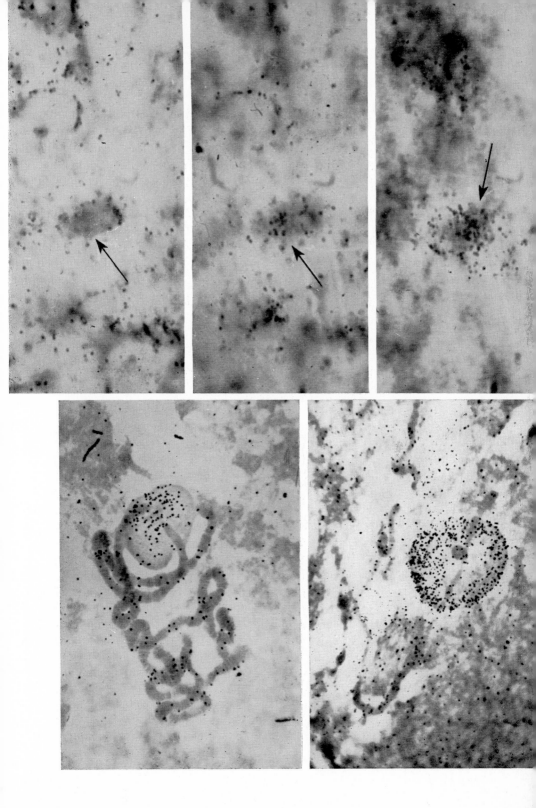

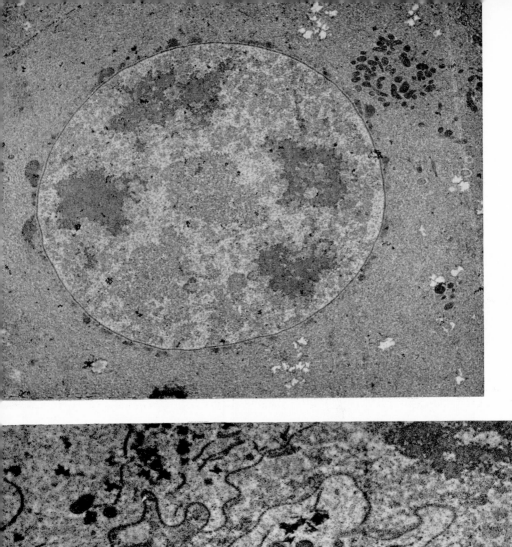
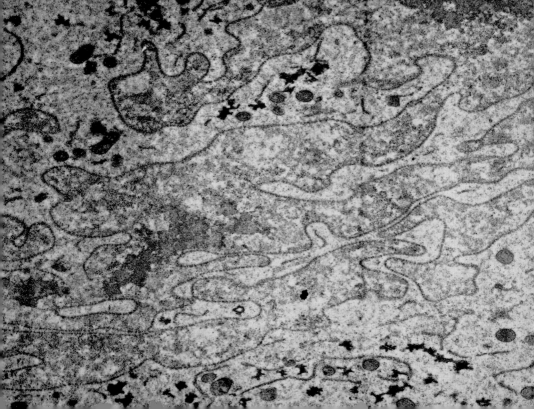

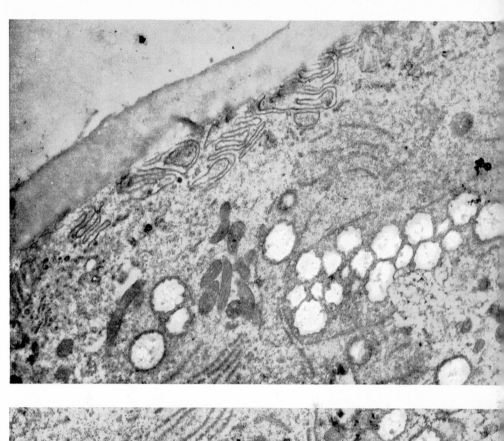

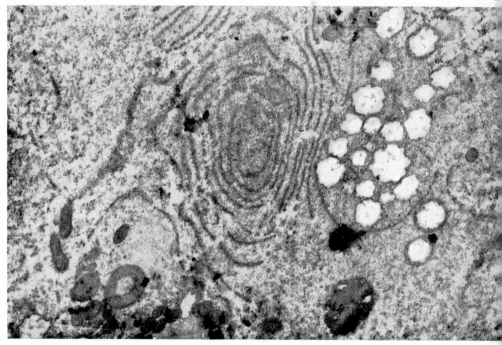

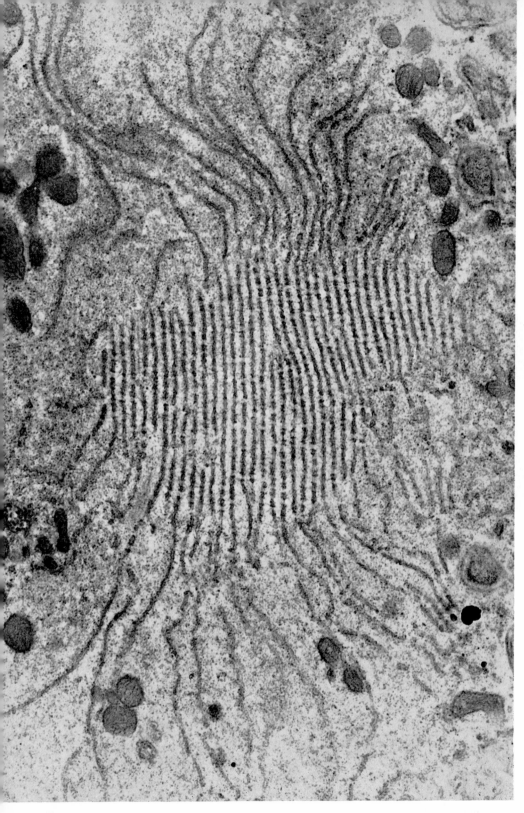

x

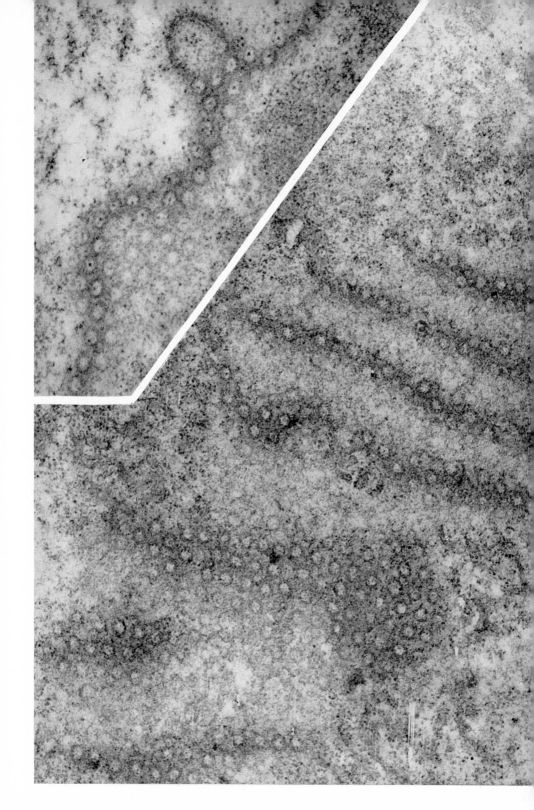

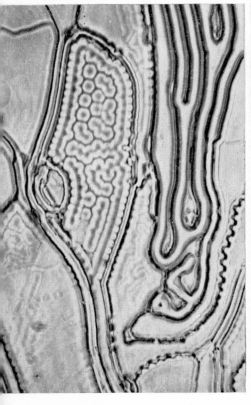

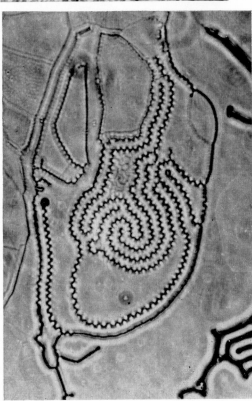

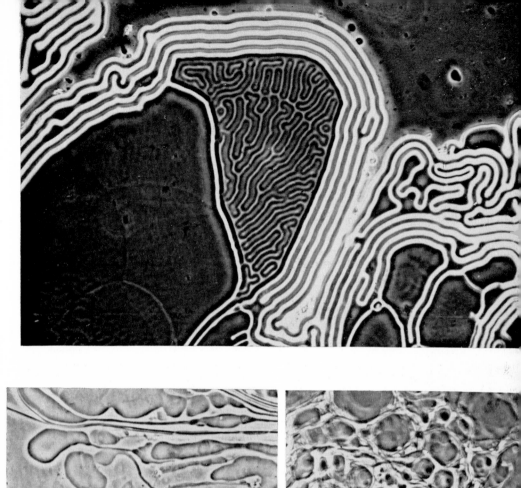

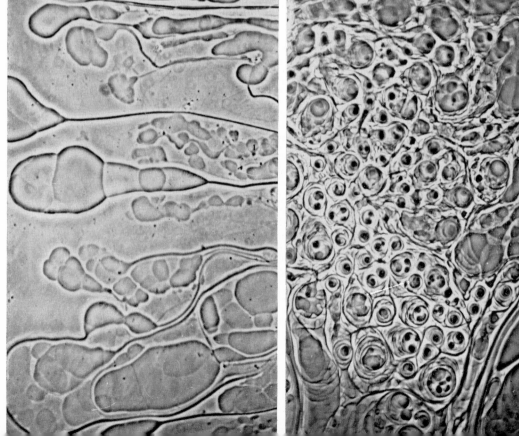

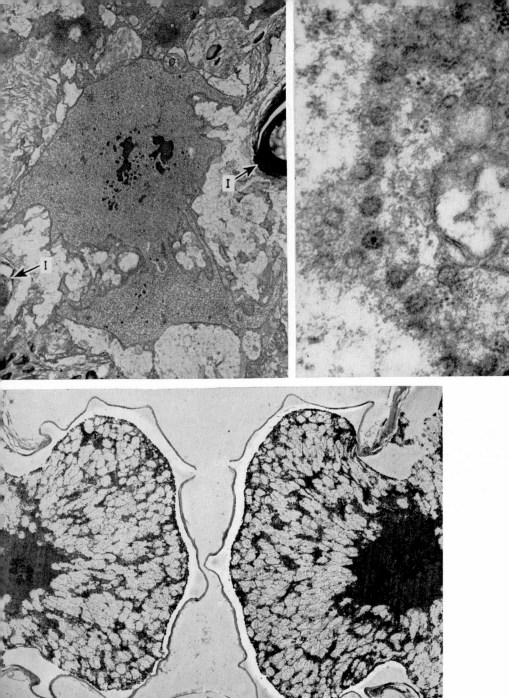

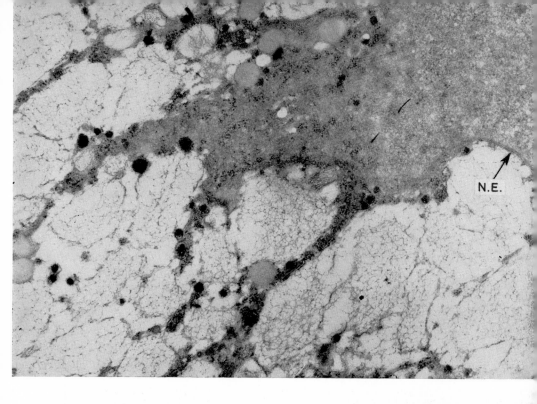

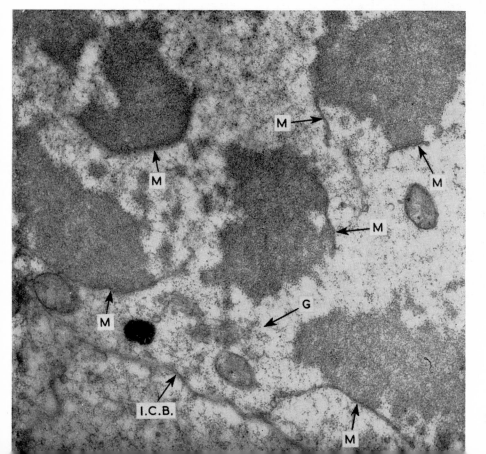

xv

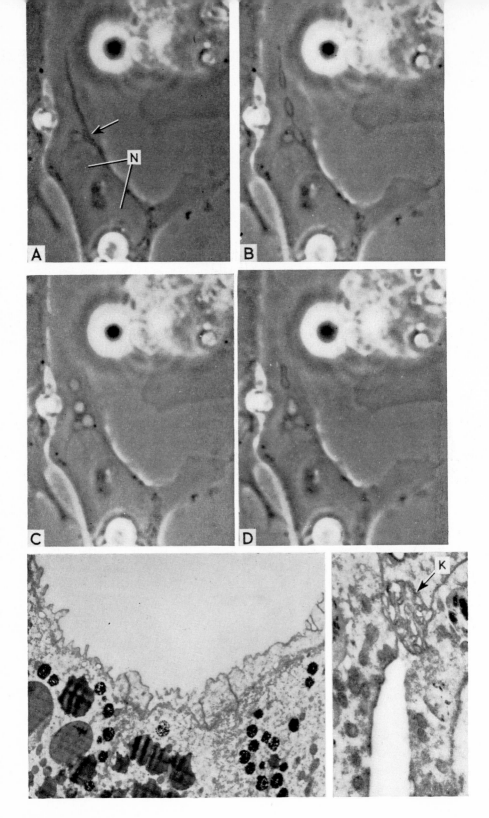

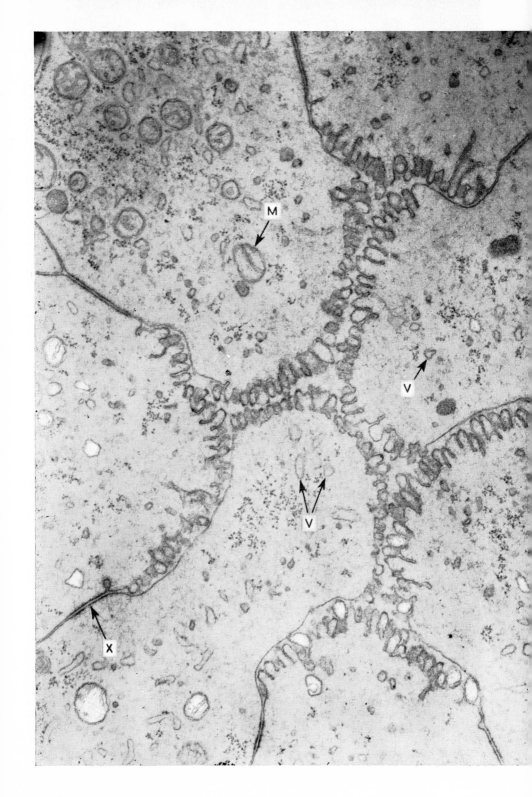

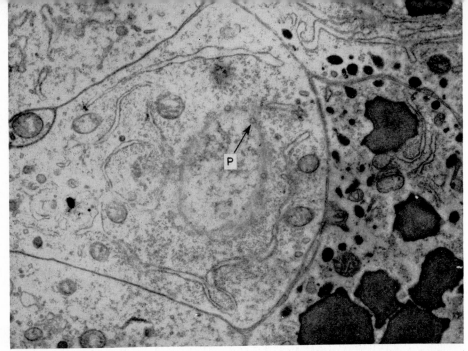

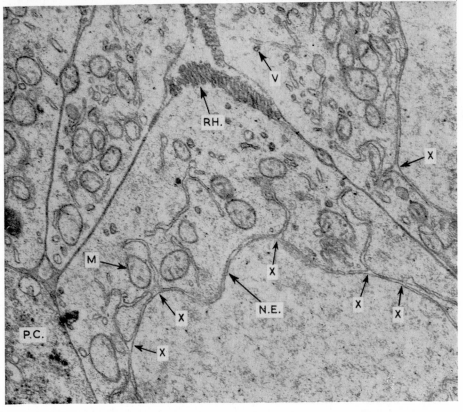

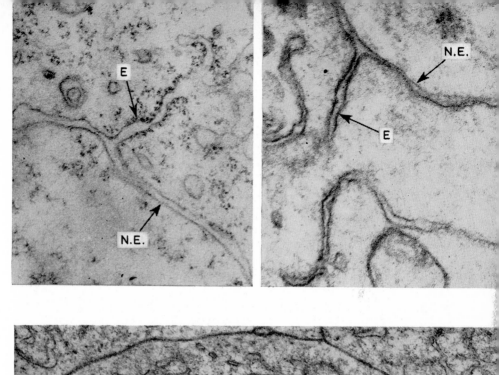

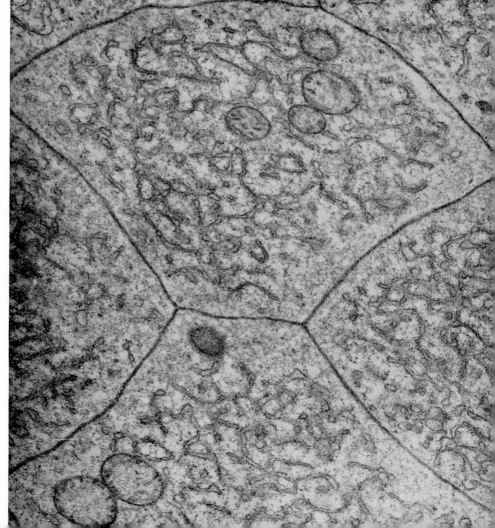

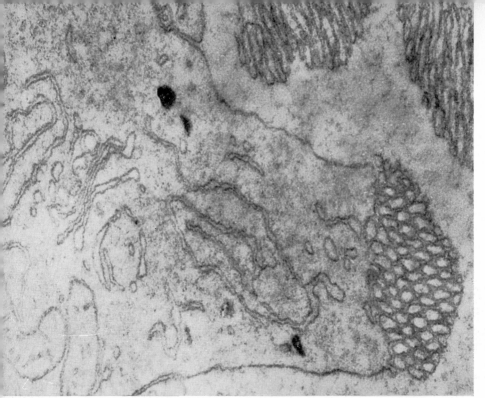

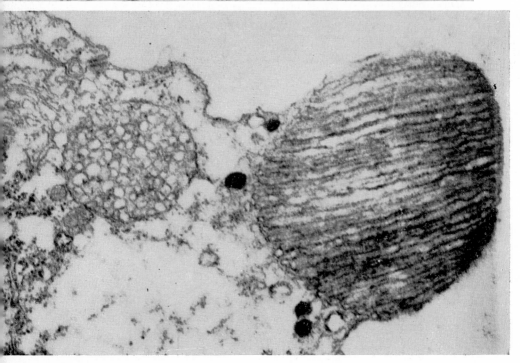

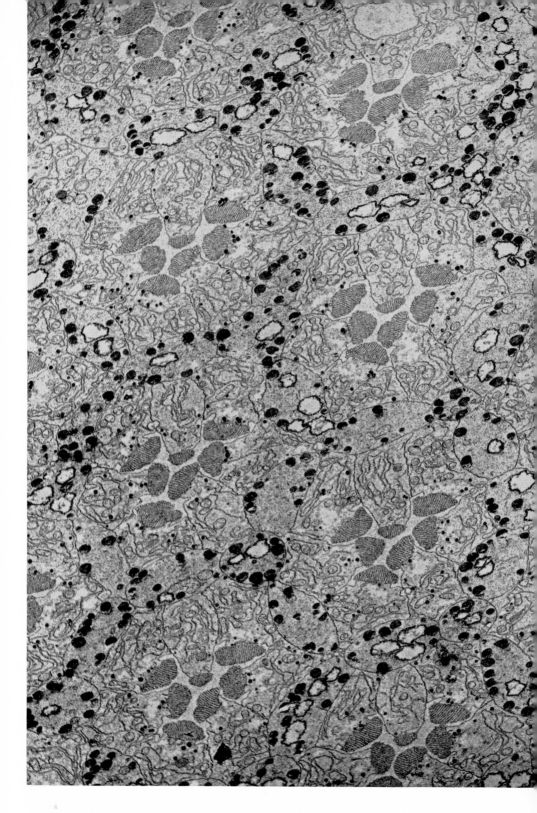

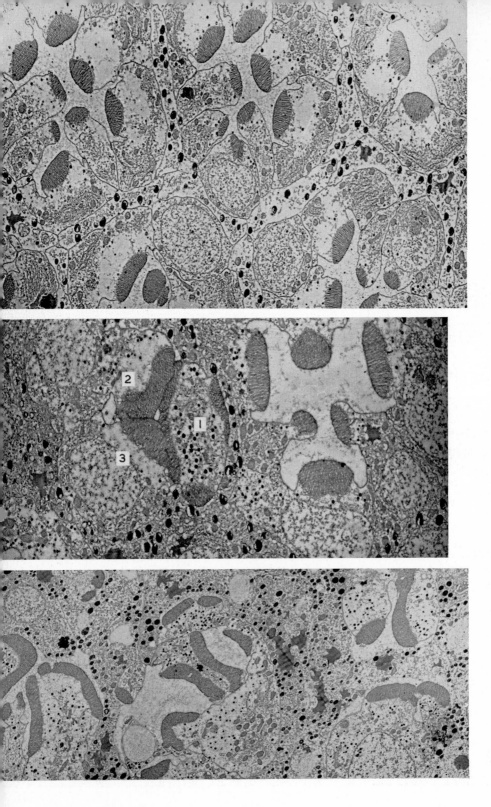

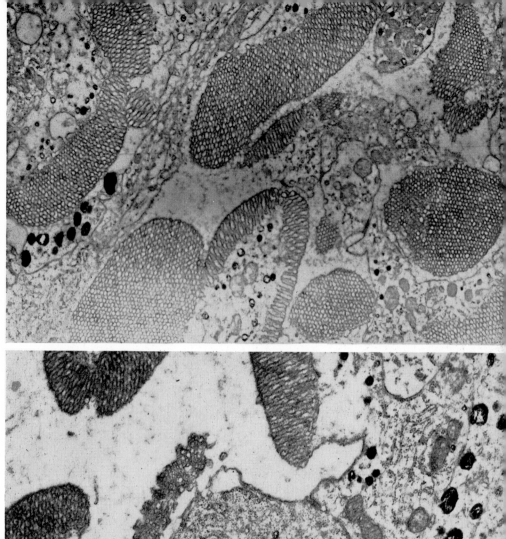
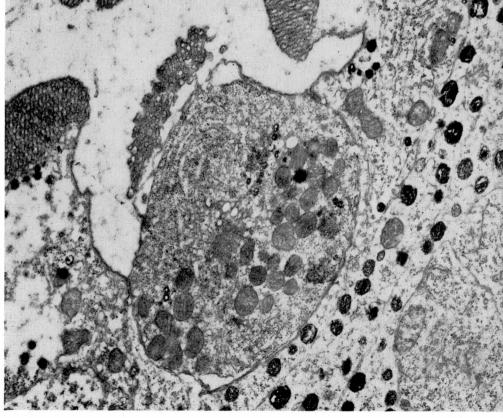

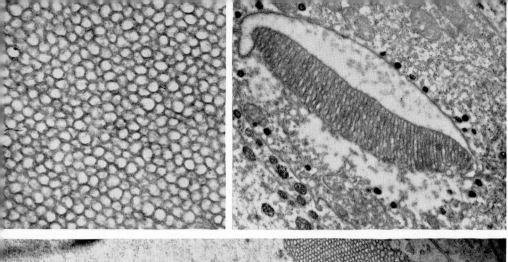

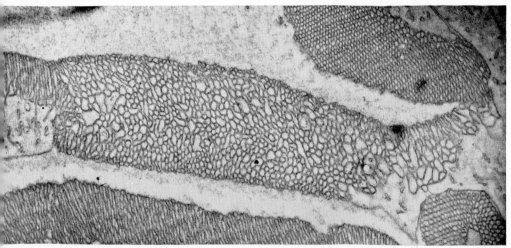

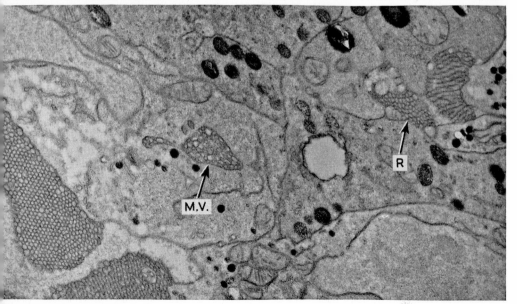

and the appearances seen in sections suggest that the main increase in rhabdomere substance is brought about by the addition of these vesicles to the inner side of the growing structure. Whether this is true or not, the point of immediate interest is that the size and arrangement of the tubular elements which make up the rhabdomeres are at first very irregular. As time proceeds the tubules gradually arrange themselves into an extremely regular meshwork. In the fully formed rhabdomere, the tubules are closely packed, so that they have hexagonal outlines, and each tubule is quite straight along its length, while they all have the same uniform transverse dimensions. The extremely regular honeycomb structure so produced must have been brought into being by the interaction of the membranes by which the tubules are lined. As we shall see later, this process can be disturbed by mutant genes.

The great complexity of structure which can be spontaneously generated by membrane systems is perhaps best illustrated by the well known myelin forms which may appear when certain lipid substances are allowed to swell in water (Nageotte, 1936; Selman, 1962). The substances which exhibit these forms best are lecithin and cephalin. The essential features of molecules which give rise to well-developed myelin forms seem to be, first, a two-pronged or tuning-forklike shape; and second, the presence of hydrophil and hydrophobe groups at the two ends. Molecules of this kind can fit together with the tuning-fork ends interdigitated and the two hydrophil ends exposed (Figure 23). Another important feature of the molecule is that the chain length should be of a suitable size so that these double molecules will lie side by side to form sheet or membrane. Such membranes will be attracted to one another by van der Waals forces and form thick piles like sheets of paper in a book. Owing to the presence of hydrophil ends of the molecules, water can enter between the sheets, so that the whole structure becomes swelled. It is this swelling which causes the formation of myelin forms.

If a fragment of solid lecithin or cephalin is placed on a slide, a drop of water added, and the preparation watched under the microscope, the myelin forms will be seen to push out from the fragment as a series of rather wormlike protrusions. Over the course of two or three days the preparation as a whole goes through a fairly definite life history. At each stage the forms, which are gradually increasing in water content as time passes, take up a number of relatively characteristic shapes. Some of these are illustrated in Plates XII and XIII, but time-lapse films give a much more vivid information of the dynamic character of these processes. At first the forms protrude more or less radially from the central fragment. It is very striking to observe the complexity and relative regularity of the arrangements which they may exhibit. We may find, for instance, long stalks, each of which bears a roughly spherical head, consisting of a ball of twisted tubular structure; other forms which are made up of a zigzag of swollen chambers of very regular size and arrangement, and also a number of other regular types. At a later stage, the elongating

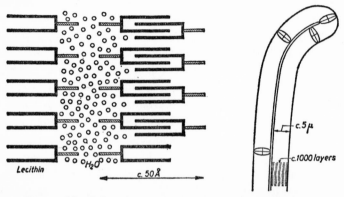

Figure 23. Submicroscopic structure of myelin forms

On the left, the tuning-fork shape of the lecithin molecule allows coupling of lipophilic parts (black), leaving the hydrophilic parts (hatched) exposed. Water enters the hydrophilic regions between layers. On the right shows a simple myelin form in the shape of a thick-walled tube, the walls having positive intrinsic birefringence (which in distilled water is reversed by negative form birefringence). The shapes of myelin forms depend on the exact distribution of water among the many elementary layers. (After Frey-Wyssling, 1948)

forms bend round and tend to run circumferentially around the mass. Again we get strikingly regular arrangements of membranes and spaces between them. By this stage the lipid is becoming highly swollen with a high water content, and there are areas which show a regular arrangement of differences in refractive index, that is to say, of regions of close or distant packing of the individual lipid sheets. These may take the form of hexagonal close packing, or various more complex periodic structures.

The main feature to which I wish to draw attention in all these different forms is the high degree of regular order which can be produced spontaneously within this very simple system. We have here some beautiful examples of the diachronic production of unit-generated forms, the units being membranes or sheets.

As a matter of fact, the simplicity may perhaps not be quite so great as appears. Myelin forms seem to be produced most readily and with greatest differentiation by relatively impure specimens of lipids. It is probable that few if any preparations of lecithin or cephalin contain only a single molecular species; certainly ordinary preparations are far from being monotypic. The diagram of the essential molecular structure of myelin forms given in Figure 23 may well be oversimplified, and something more may be involved than the interlocking of tuning-fork shapes. Certainly some double-pronged molecules fail completely to form typical figures; for instance, a synthetic lecithin, in which the aliphatic chains were saturated, which Selman has recently studied in our laboratory. It is known that in some figure-forming preparations, for example, mixtures of lecithin and cholesterol, complex-formation between the components plays an important role (Dervichian, 1958), and it may be of fundamental importance in all or most cases in which elaborate figures appear. It is probably also essential that the substances involved should produce liquid, and not solid, monolayers on the surface of aqueous media.

Rather few attempts seem to have been made to study the form production which can be elicited by mixing together known quantities of definite substances. In general, it has been found that the addition of protein to the watery solution in which the lipid is placed tends to restrict the appearance of myelin forms. It seems rather likely, however, that this would not necessarily be the case for all combinations of lipid and protein. For instance, one of the most interesting of such forms was produced by Crile, Telkes, and Rowland (1932) in a system which consisted of "crude brain lipid" mixed with "crude brain protein." They obtained small spheres, each of which enclosed a smaller internal globule having roughly the appearance of a nucleus, while the external surface was drawn out to a large number of fine filaments. The exact mode of formation of such structures is very little understood. In general terms it might be expected that suitable protein molecules might be adsorbed onto the lipid membranes in such a way as to stabilize them, and that more complex and less transitory arrangements could then be produced (Figure 24).

In the present context, the interest of the myelin forms is that they provide an example of the production of structural regularity by means of the interaction of sheets in a relatively simple system. They are, of course, also of great interest in another way, as providing models of various aspects of cell behavior. For instance, amphibian embryonic cells can in certain types of solution be brought to throw out fine pseudopodia, so that they look extremely like the globules of mixed brain lipid and protein mentioned above. Holtfreter (1948) has also discussed in considerable detail the parallelism between the more usual pseudopodial

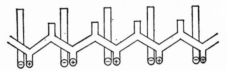

Figure 24. Two types of lipoprotein film

The zigzag polypeptide chain at the top is attached by relatively weak bonds to the lower surface of a lipid film on an aqueous medium. Adsorption is reversible. In the lower diagram, the protein penetrates the lipid film and becomes irreversibly adsorbed. (From Matalon and Schulman, 1949)

behavior of amphibian cells and myelin forms. There is little doubt, indeed, that in such comparisons we are not dealing merely with a parallelism, but with a real similarity in underlying structure (cf. Danielli, 1958). The plasma membrane contains a large proportion of lipid, and its behavior almost certainly has a real similarity to that of a myelin form, although in the latter the motive force producing the changes in form is the inhibition of water; and this can scarcely be the case in the living cell.

Although the mere fact that regular structures are produced in this way is enough to give us confidence that membrane systems are one of the types of forming-producing mechanisms to which we can appeal to explain certain biological structures, we should not be content with this. We should like to understand the exact processes by which these orderly arrangements are generated. For certain of the structures mere inspection, particularly of time-lapse films of their formation, is enough to allow us to offer fairly plausible guesses. For instance, it is apparent that an elongating myelin tube tends to run along the surface of any pre-existing myelin structure in the neighborhood; and this is, perhaps, sufficient to account for some of the spiral arrangements of tubes which we see, such as those in Plate XII. Again, one can understand how a zigzag arrangement of chambers, once it got started, might be expected to continue to grow in the same form; as each chamber swells up it gradually produces a sharp re-entrant angle between it and the pre-existing structure, and at such a re-entrant angle there will be special stresses which might well

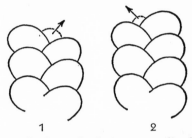

1   2

Figure 25. Growth of zigzag myelin figure

If such a figure tends to grow most rapidly at a re-entrant angle, it will pass from the stage shown at 1, when a protrusion is being pushed out to the right, to the stage in 2, when a new protrusion starts being formed toward the left.

cause the next phase of enlargement to take place in this region (Figure 25). It seems rather likely that, in any region of a cylindrical tube which is growing rapidly in length, the constituent membranes would be sliding over one another; and this might make them more permeable, so that their tendency to expansion would be further increased: possibly this is the explanation of the long thin stalks with contorted heads.

Very little detailed study has been devoted to such questions. Holtfreter (1948) described some interesting experiments in which granules or droplets of particular substances, for instance, fatty acids, were introduced locally into myelin forms and shown to have effects on their later development. However, this work did not go far enough to throw much light on the production of regular structures by such systems, which is one of the most striking, and for the general theory of morphogenesis, one of the most important aspects of the whole matter. Holtfreter's work

has not been followed up, but an experimental study of the morphogenesis of orderly structures in mono- or polytypic myelin systems would
seem likely to be very rewarding to anyone who wishes to help bridge
the gulf which now separates the macromolecular from the biological.

Instruction-Generated Forms[1]

It would be possible for an orderly structure to be produced by a system which consisted of a number of units, either of one or several kinds,
together with a set of instructions as to how they should be assembled.
Such a system would differ from a unit-generated one, of the kind considered previously, in that in the latter the manner in which the units
become joined is dependent on their intrinsic nature; whereas, in the
category now under consideration, it is determined by a set of instructions external to the units themselves.

Instructional systems are, of course, those commonly used by man-
kind. They are exemplified, for instance, when a bricklayer builds a
house according to the instructions given him by an architect. Here
the architect's working drawings amount to a series of instructions according to which the bricks are assembled, by a diachronic process, into
the finished structure. It is indeed easy to imagine an instructional system
operating by means of diachronic processes, although there is no theoretical reason why a set of instructions should not in some cases be carried
out simultaneously by a synchronic process. We might, for instance, describe in this way (i.e., as a synchronic instructional system) what happens when the sergeant major shouts "Fall in" to a group of dispersed
soldiers.

In the context of the formation of biological structures, the main
attractiveness of the idea of instructional systems arises, perhaps, from
a sort of carry-over from the way in which information theory is commonly applied. We have become used to the statement that genes carry
information. "Instructions" are after all a kind of information. Might
we not suppose, therefore, that the genes carry instructions as to how
the elements of the living system are to be assembled into structures?

---

[1] This category was not discussed at the original lectures, and I should like to
express my thanks to Dr. Ruth Sager for drawing my attention to it in a discussion
following the lecture.

The difficulty, however, is to see what physical form these instructions could take. If, for instance, actin and myosin do not, on account of their own inherent properties, unite together in a particular way (in which case they would constitute a unit-generated system) how can any gene provide instructions that they shall be united in that manner? One way in which this could be done would be for the instruction-giving gene to control the environmental conditions in such a way as to alter the properties of certain reactive sites on the actin and myosin molecules, so that they tended to enter into specific physical relations with one another. It might be possible, for instance, by causing a gradual change of pH, to determine a sequence of interactions between protein molecules by which a certain structure would be generated, and we would then have an example of a diachronic instructional system. But it is not easy to see how any great amount of detail could be incorporated in any system of this kind which might plausibly be supposed to operate in biological organisms. Instructions conveyed only by gradual changes in the general environmental situation could only be of a very general kind. In so far as the structures involved a large amount of particular detail, this would have to be in major part incorporated in the nature of the units; the instructions could only be operating as comparatively minor modifications of an essentially unit-generated system.

So far as I can see, the only other practicable way of incorporating a set of instructions into a biological system is to embody it in a material structure, in the immediate proximity of which the new structure is to be formed. This is as though an architect could not hand to a bricklayer a scale drawing of what was to be erected, but had to peg it out full size on the ground. Systems such as this make up a very special class of the general category of instructional systems. They are, in fact, a class which, because of its great importance in biological systems and its special character, is worthy of a name of its own. I have referred to them as "template systems," and they form the subject matter of the next section.

Template-Generated Forms

In systems of this kind the new form is produced by the assembling of a number of units in the immediate neighborhood of some pre-existing structure whose pattern determines their orderly arrangement. In such cases the pre-existing structure is said to act as a template.

There are two major types of template action. First, those in which the elements of the template pattern come into operation one after another (diachronic templates); and second, those in which the whole template acts simultaneously (synchronic templates). The former category is probably of rather little importance and we can dispose of it immediately. It provides an interesting theoretical link with the instructional systems considered in the last section, since we could regard the various elements in the template pattern which come into operation one after another as a sequential set of instructions. An example on the molecular level would be the sequence of genes, investigated by Demerec and others (1956) in Salmonella, which operate on successive steps in the transformation of certain metabolic syntheses. For instance, four cistrons concerned in the synthesis of tryptophane are arranged along the chromosome in the order of the steps in the chemical reactions which their enzymes catalyze. The same arrangement is found with several cistrons concerned with histidine, and there are still other examples of the same kind, although orderly sequences of this sort are by no means the general rule.

One could say that the set of genes provides a diachronic template by which the structure of the final molecule is brought into being. At levels higher than the molecular not many examples of such systems can be found if only because it is unusual for one part of a living organism to be passed along the surface of another. However, a few exist; for instance, the structure of a bird's egg at the time it is laid has been brought about by its passage over a definite sequence of regions in the oviduct at which the various layers of albumen and shell have been secreted.

In the majority of template actions with which we are concerned in biology, the various elements in the template pattern probably act simultaneously or synchronically. This does not, of course, necessarily imply that the reaction to the template should cease within a short time, leaving the structure in the form in which it first appears. A synchronically acting template may, indeed, produce a synchronically formed structure, as it presumably does, for instance, in gene reproduction; but we must also be prepared to find a synchronically acting template which produces a diachronically elaborated structure, that is to say, one which goes on developing for a considerable time through several stages before reaching its final form. We shall come across many examples of this in the more complex multicellular entities. A clear example in a unicellular form is provided by the development of the new half cells in Micrasterias, described in the next chapter.

The most important subdivision of the class of synchronically acting templates is not according to whether they control the formation of structures which are synchronically or diachronically elaborated. A more important point is whether these structures are, or eventually become, exact copies, or at least very simply coded replicas of the templates, or alternatively whether they are something quite different to the template, which could be related to it only by a complicated coding relation. The examples to be discussed below, will, it is hoped, make this distinction clear.

a. Copying or simple coding. Perhaps the most clear cut example of exact copying is the process of DNA replication. Although the details of this process are still obscure, it is clear enough that the original DNA molecule acts as template on which an identical duplicate of itself is eventually produced. At various stages in the existence of such DNA molecules, certain molecules of protein and RNA are also, it is thought, produced on the same DNA template. The RNA, and probably the protein formed in this way, are thought also to be very specifically related in structure to the RNA; that is to say, although obviously not an exact copy, they are related to the template by some simple coding relation.

On a supramolecular level the addition of new laminae to an annulated laminar stack would seem to be a good example of the formation of an exact copy on a template. If, as has been suggested, the stacks originate in the neighborhood of the nuclear envelope, the formation of the first lamina may also be an exact copying process, or one involving a simple coding relation. Similarly, if the ergastoplasm in general is formed from the nuclear envelope, its production is also a process of the same general type.

The growth of the nuclear envelope itself provides an example of a particular kind of exact copying template action which seems of frequent occurrence in biological systems. The nuclear envelope has a complex structure involving two electron dense layers and a number of annuli or pores. We can scarcely doubt that the existing envelope acts as a template for the formation of new material as the envelope increases in area.

However, this new material is not found lying side-by-side with the exist-ing envelope, but becomes incorporated into it by some sort of intussus-ception. The exact mechanism of such intussusceptive template action remains obscure, but there is much evidence which suggests that such processes are rather common. For instance, mitochondria are relatively complex structures formed from membranes on which enzymes seem to be arranged in rather definite order; it seems probable that the mem-branes must be capable of expansion in area by intussusceptive template action.

If we are willing to think in a sufficiently abstract way, we can find examples of intussusceptive action even in highly complex multicellular systems. For instance, the growth of a segmented worm, to which new segments are added in a proliferating zone, could be regarded as an example. However, in such cases the processes by which a new segment comes into being and becomes elaborated are so complex that there seems little point in drawing attention to the fact that the over-all effect falls logically into this category. It would be biologically more meaning-ful to try to analyze the process of segment formation into more ele-mentary processes; for instance, those by which the particular organs in the segments are produced. If one could do this one would probably find oneself confronted with many examples of template action, but the situa-tion would undoubtedly be more complex than the over-all result might lead one to suppose. We might quite likely find that intestine was formed in the new segment even if the intestine had been removed from the nearby segments which were acting as the templates. We should, in fact, probably find that we were dealing with examples of the template pro-duction of noncopies, a category of actions which we must now consider.

b. Template production of noncopies. Actions of this kind are exceed-ingly common in multicellular systems. The category, in fact, covers the whole range of what is usually referred to as the "induction of organiza-tion," or as I have called it, the "induction of individuation." When a part of the archenteron roof is implanted into a newt gastrula and induces a well-formed part of the nervous system conformable to it, then the im-planted organizer is acting as a template which induces a noncopy. Similarly, when competent ectoderm is grafted into the ventral head region of a tadpole, it differentiates into a mouth and other head structures and the positions of these are determined by the underlying materials acting as templates (see Chapter 5). Again, in the regeneration of an amputated vertebrate limb, the stump acts as a template which determines that the regenerate shall form, not a replica of the stump, but that part of the limb which was removed. Other examples of such templates which produce noncopies are the conditions which Stern (1954) refers to as "pre-patterns," that is, situations existing within the body of a developing Drosophila, which determine that any tissue competent to form bristles, sex-combs, etc. will do so at particular places. We shall return to these also in Chapter 5.

In all these cases the form of the new structure is determined, to some extent at least, by the nature of an already existing structure, but the new thing that is formed is not a copy of the old. Moreover, in none of these cases does it seem plausible to suppose that the coding relation between the old and the new is as simple as that between, say, DNA and the RNA corresponding to it. In many cases indeed we know that the coding relation must be exceedingly complex. For instance, if the femur is removed from a newt limb, the boneless stump will, nevertheless, regenerate the missing part of the limb, complete with its bony skeleton. For our present purposes I do not wish to attempt to discuss the voluminous, complicated, and perhaps not very illuminating data, which have been accumulated about the relations between the template and its production (or in more usual terms, between inducer and induced) in such situations. All I wish to do at this point is to mention that such processes form a very common category of structure formation in biological systems.

Condition-Generated Forms

These are forms which arise, not from the nature of the units out of which they are built, or from the action of a set of instructions or a template acting on these units, but rather from the interactions of a number of initial spatially distributed conditions. It is, perhaps, rather difficult to envisage what is meant by this abstract definition without the aid of a concrete example. I will, therefore, describe a very simple system in which a number of conditions generate a definite geometrical pattern.

Imagine a roughly cylindrical mass of tissue as it might be a developing limb. Suppose that within it there are the following initial conditions:

(1) there are precartilage cells which tend to collect together along the axis of the cylinder; (2) when the concentration of these cells increases beyond some limit, they degenerate and disappear; (3) the threshold for this degeneration decreases in the proximo-distal direction. Then in the proximal region, in which the threshold is high, the precartilage cells continue to accumulate and to form a massive axial cartilage. Further distally the congregating cells over top the limit and start to degenerate. If the mass of tissue is not quite cylindrical, but flattened in cross-section, this process would split the central accumulation of cells into two in the distal regions. More distally still, each of these two condensations splits again into two further subdivisions. The initial conditions, in fact, generate a structure consisting of one condensation proximally, followed by two more distally, followed by four further distally still.

A simple model of such a system can be constructed as follows: A large sheet of chromatography paper is sprayed with mineral acid and alkali in such a way as to set up a uniform pH gradient, so that the paper is acid at the top and alkaline at the bottom. A simple air driven spray gun is now filled with a mildly alkaline solution which contains an indicator which is colored in acid pH and colorless in alkaline. The paper is held in a vertical position with the acid at the top, and a single jet of the spray gun is run vertically down it from top to bottom. The gun delivers a jet in which the quantity of alkaline fluid is greatest in the mid line and falls off toward the sides. At the acid region of the paper, a considerable width of the solution jetted onto the paper becomes colored. At some point farther down the paper, the alkali in the central portion of the jet combined with the solution in the paper just suffices to decolorize the indicator; from this point downward the colored bar splits into two portions. We find ourselves therefore with the paper showing a single colored bar at the top, which splits into two down farther (Figure 26).

In both these cases, the imaginary biological model and the actual experimental model, we have finished up with a geometrical pattern which was not incorporated in any one of the general conditions with which we started, but which is engendered wholly by their interaction. All the initial conditions, in fact, were pervasive ones; that is to say, although they were not homogeneous over the whole field, nevertheless, they varied in a uniform manner over the whole of it. The boundaries of the structures which appeared showed the loci at which the values of the different variables involved had some definite relations to one another. It is to systems in which structures are produced in this way that I wish to refer by the phrase "condition-generated systems."

There are two main subclasses of such systems: those in which the conditions involved are subject to stochastic variation and those in which they are determinate. Form generation by determinate conditions has many affinities with the template generation of noncopies in that both depend on pre-existing spatial arrangements which are not precisely the

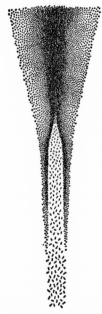

Figure 26. Condition-generated pattern

same as the form eventually produced, but differ in that the latter involves processes of induction, that is, interaction between the region in which the form develops and the template; whereas, in the former the conditions are to be thought of as pervading the whole mass.

a. Stochastic conditions. The possibility of such systems was probably first suggested by the mathematician Turing (1952). He seems to have had a somewhat scanty knowledge of biology, and like many of those unfamiliar with the facts of embryology, he thought that its problems, which are difficult enough in all conscience, were even harder than they actually are. Just as some geneticists seem to feel called upon to explain differentiation into alternative end-states by purely genetical processes which do not involve the regionally differentiated cytoplasm of the egg, Turing felt that we have to be able to explain the appearance of regular formed structures from a completely homogeneous initial situation. He envisaged the typical blastula, for instance, as a radially symmetrical spherical figure, and seems to have been unaware that it possesses both an animal vegetative axis and a dorsal meridian. However, he proceeded to show how, starting from this very unpromising beginning, much more definite structures could appear than seems likely at first sight. The only account which he lived to give of his ideas is highly mathematical, and his development of them was at an admittedly early stage at the time of his premature death. The details—in fact I must confess even the majority—of his exposition is beyond my mathematical comprehension. I think, however, that the basic lines of his thought can be made clear even to the simple biologist.

Turing starts by considering an area—the whole treatment is in terms of two dimensions for greater simplicity—in which there are a number of chemical substances (morphogens) of which the final structures are composed. These morphogens both react together and are also capable of diffusing from place to place within the area. In a determinate or nonstochastic system, the concentration of each one of the morphogens would eventually settle down to some equilibrium value, which would be the same over the whole area and would depend fundamentally on the rate constants of the interactions by which the morphogens generate or destroy each other.

Turing's treatment consists of a mathematical examination of the various types of instability which may arise when the system is slightly disturbed from its equilibrium state by random disturbances. If at a particular place the concentration of a certain morphogen is by chance raised above its equilibrium value, this will not only cause an alteration in the amount of it which reacts with other morphogens, but also the morphogen will tend to diffuse away from this region. Turing treats the situation first for the simplified case of a circular ring of tissue; this is in effect a way of getting rid of the need to consider the boundaries of the area, which can be treated as a closed domain. He found it possible to solve the equations and to give the values which the concentrations of the morphogens asymptotically approach as time passes after an initial disturbance—the more actual case of continuing random disturbances was not dealt with in his first paper.

From the biological point of view, the most interesting points that emerged were that under certain conditions (i.e., with certain relative values of rate constants, diffusion constants, etc.) some of the initially disturbed points will become reinforced at the expense of others. In this way a region in which the concentration of a certain morphogen has been slightly raised may build up to a larger area in which the concentration

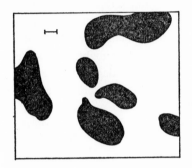 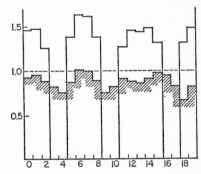

Figure 27. Turing's patterns

On the left, a dappled pattern developed in a two morphogen system with marker of the unit length used in the calculations. On the right, the gradual build-up of a periodic pattern is a one-dimensional Turing system, for example, a row of cells. The original homogeneous equilibrium is indicated by the dotted line; the line with hatchings, after a certain length of time; the continuous line, the final equilibrium reached. (From Turing, 1952)

is considerably higher than it had been, while in other areas the concentration of the same morphogen will fall to quite a low value. The areas of high concentration are usually rather irregularly disposed, forming what one might call a dappled or randomly spotted pattern (Figure 27).

The most interesting of all Turing's results, however, was the demonstration that under certain conditions the pattern appears as a periodic system, that is to say, the areas of high concentration form a series of stripes of relatively uniform width and with a fairly characteristic distance between them. The wave length of these periodic patterns is determined by the particular values of the reaction rates and the diffusion constants. It is a chemically determined wave length (Figure 27).

It is, I think, a major contribution to theoretical biology to show that any regular pattern at all can emerge from a system which consists initially only of a homogeneous expanse disturbed by purely random processes. But once we have paid the tribute to Turing's work, which it very well deserves, we need to go on to consider the points in which it is still deficient. This was, after all, only a beginning which Turing himself would certainly have liked to carry much further. I think the two points which a biologist would have liked immediately to raise with him would be these. First, the patterns produced are rather irregular. The computer, which was calculating for him the situation at successive times, produced a number of patterns which were either irregular dapplings, or at best, roughly periodic stripings. How much would such patterns be tidied up if one could introduce into the system some element of whole control, some feedback from the already existing area? Presumably any system which involves diffusion contains already a certain degree of feedback; but could one not postulate some other physical system in which a greater intensity of feedback was in operation, and would one not in this way generate something a good deal more regular and orderly than Turing produced?

The second, and perhaps more important, point concerns the chemical determination of the wave length of periodic structures. In any particular system the chemical rate constants would presumably remain constant as time passed, so that the wave length would also remain constant. In most biological systems, on the other hand, the wave length of periodic structures is related to the overall size. For instance, Turing points out that in his circular ring of tissue, chemical systems might generate a number of high points, at which tentacles might form. But, according to his system there should be more tentacles in a large cylindrical organism than in a small one, since the distance between the tentacles would remain constant and not depend on the total circumference. However, in living coelenterates the number of tentacles is about the same in large and small specimens, and the distance between adjacent tentacles does depend on the total circumference. It is true that certain constant-period systems are known in animals (see the example of *Theodoxus fluviatilis* in Chapter 6, Figure 55) but they are certainly not the only, or even the most common, type of periodic structures in the biological world.

To the nonmathematician there appear to be several modifications of the Turing system which seem likely to make the periodicities less dependent on the unaltering chemical rates, and more plausible in connection with those we meet in living organisms. For instance, Turing considered only the cases of an infinite plane or a continuous ring. If we could deal with a bounded area, outward diffusion of morphogens across the boundary would modify the wave length and would presumably make it somewhat more responsive to tissue size. Again, in the case of the ring, where Turing was able to develop the best defined wave length, each high point interacts only with the two highs on either side of it. But consider a rectangular area in which there were four highs, one at each corner; this is a disposition somewhat similar to the scutellar bristles in Drosophila, for instance (Figure 28). The high point at corner one will inter-

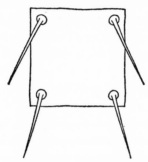

Figure 28. Pattern with hairs at corners

act not only with those at the two corners nearest to it (corners two and four) but also with the diagonally opposite corner, three. If the size of the square of tissue becomes so small that the chemically defined wave length does not fit easily between corners one and two, or one and four, it may come close to a decent fit with the distance between one and three. In this way the arrangement of the highs into a pattern at the four corners of a square might represent the best attainable fit even if the square became quite small in comparison with the wave length. However, these are only suggestions which seem intuitively to have some plausibility. It is clear that much further precise mathematical work is called for on the lines opened up by Turing before we can really decide how powerful is the form-generating mechanism to which he has drawn attention.

Maynard Smith (1960) has made the interesting point that any pattern that arises from such stochastic mechanisms (and indeed patterns arising from somewhat more determinate processes) is bound to show considerable phenotypic variance. From an inspection of the amount of variance commonly found in anatomical characters, he comes to the conclusion that it is not plausible to appeal to such mechanisms to explain the generation of series of periodic elements which number, precisely, more than about five or six at the most. When the number of repeated elements is more than this, the series ought to be variable in the numbers it contains in different individuals. There are many instances in which variation does not appear. For instance, many species of worms, myriapods, and other segmented animals have quite a high number of segments (say, between twenty and forty) which show extremely little variation among individuals of a species. Maynard Smith suggests two ways in which such high but relatively invariant numbers could be produced. They might, in the first place, result from two successive phases of segmentation. A thirty-segment animal, for instance, might pass through an early development stage of only five segments, and then each of these five is later divided into six. A second possibility is that the segments, although similar in general appearance, are not in fact identical, but each have some specific, presumably chemical, identity. If this were so, the general process which generates the periodic structures might be controlled by some form of chemical counting, which would bring the series to an end when the appropriate chemical constitution had been attained.

The suggestion of chemical counting invokes, on theoretical grounds, the idea that similar looking biological structures may actually differ from one another in some usually unknown chemical way. There is a good deal of somewhat indirect evidence that suggests that such phenomena are in fact very common in practice. For instance, tissue culture studies on the limbs of birds have shown that the different bones acquire at a very early stage specific individual characteristics which must presumably be chemical in nature; for instance, the humerus, radius, and ulna, while still in the stage of cartilagenous rudiments demand slightly different synthetic media for optimum growth *in vitro* (Abbott and Kieny, 1961). Again, the fact that genes can be found which specifically remove particular bristles from the thorax of Drosophila, suggests that each of these bristles has a certain particular individual character. We shall return to this topic later. It is mentioned here only to make the point that the appeal to a process of chemical counting is by no means biologically unreasonable.

b. Determinate conditions. Presumably this is the most common type of form generation, at least at the level of the microscopic and upward. Most organisms develop from eggs which have some structure in the form of localized special plasmas such as the amphibian gray crescent. We probably do not in practice often have to deal with completely homogeneous systems in which one can appeal only to stochastic processes to generate form as Turing did. Most form generation, at the cellular and tissue level, at least, starts from some determinate arrangement of conditions, which acts somewhat like a template which produces a noncopy of itself.

One of the most convincing types of evidence that a form is condition-generated is provided when it can be shown that profound modifications in it can be produced experimentally by a change in general conditions. In some instances a mere alteration of the osmotic strength, or composition of the medium in which an organism is living, can produce an astonishingly large transformation of it. A remarkable example is provided by the flagellate Naegleria (Figure 39), and another equally striking case is the change in the color pattern on the shell of *Theodoxa fluviatilis*. In neither case does a knowledge of how to control the switching from one pattern to another lead to any other deeper insight into the nature of the patterns themselves. Attention has already been drawn above to the difficulty of guessing at the nature of the conditions which engender many types of biological form which we do not know how to alter by experimental means.

Even though we cannot usually identify the conditions by which various forms are produced, there are some general points to be made about them. Condition-generated forms must usually be extremely polytypic, that is to say, very many conditions of different natures must be involved in their production. This is perhaps best demonstrated by the facts of genetics. In organisms whose hereditary material has been thoroughly investigated, it is common to find that the form of even relatively simple structures may be affected by very large numbers of different loci. For instance, several tens of factors are known which affect the relatively simple over-all shape and venation pattern of the wings of

Drosophila. Each of these genes, or rather each gene product, would correspond to a "morphogen" in the sense of Turing. It seems probable that most multicellular structures are similar to Drosophila wings in that the shape is affected by very many different factors. In such polytypic systems there usually seem to be powerful feedback interactions between the various components. This is demonstrated by the evidence for what has been called "canalization." Canalization implies that the developmental processes leading to the formation of a structure or a tissue, though of course not entirely inflexible, offer some resistance to the effects of agencies which tend to disturb them and to bring about an abnormal end result. The effect of this is that the normal adult phenotypic is to some extent buffered, or protected by thresholds or quasi thresholds against disturbing influences. An example is the pattern of venation in the region of the posterior cross-vein in a Drosophila wing (Waddington, 1955). For the investigation of this, a number of stocks were built up, in some of which parts of the normal cross-vein were absent, while in others additional pieces of vein were present. In all these stocks the differences from normality were brought about by the presence of many genes, each of slight effect, not by one or two major genes. When a number of crosses, back-crosses, etc., among the various stocks were carried out, it became clear that the normal wild-type pattern of venation could be produced by quite a large range of different genotypes. A normal-appearing individual can in fact contain quite a high dose of vein reducing or vein increasing genes. In contrast to this, once the thresholds of gene dosage are over stepped and phenotypic abnormality is produced, the degree of abnormality provides quite a good guide to the actual genetic constitution of the individual concerned. It is only the normal pattern which is well buffered and protected by considerable thresholds. Some further examples of such canalization of structures and patterns will be discussed in the last chapter.

It must, of course, be realized that many of the processes of form generation in the biological world are of mixed type involving processes from more than one of the categories described above. For instance, in some experiments of Grobstein (1955), the epithelial cells of various glands were grown in culture and took the form of a simple sheet, unless they were acted on by substances diffusing from appropriate mesodermal cells, in which case they became transformed into tubular structures.

The tubules only appeared when the epithelial portion of a gland was acted on by substances coming from the mesodermal components of the same gland. Thus, the appearance of tubular structures was dependent both on the inherent nature of the epithelial cells (and was to that extent unit-generated) and also on some influence other than the cell's own character; and in so far as an influence over and above the nature of the epithelial cells is operating, the forms produced are either instruction-generated or condition-generated. Grobstein's formulation, in which he speaks of the influence of the substances diffusing from the mesoderm as an inductive action, suggests, at first sight at least, an interpretation in terms of instructions. On the other hand, Weiss (1959) stresses the point that there is no need to invoke anything more elaborate than condition-generation. In a later chapter, dealing with other examples of embryonic induction we shall discuss several other cases in which unit-generation and condition-generation are both simultaneously involved.

Conclusion

I have tried to show how we can give an abstract description of a number of different types of form generation: generation by units, by instructions, by templates, and by conditions. These major categories and their subcategories provide quite a battery of mechanisms to which we can appeal to explain the appearance of form in biological structures; and we can, perhaps, when confronted with any structure, usually give a more or less plausible guess as to which category of process has occurred. In nearly all cases, however, it is still extremely difficult, if not impossible, to proceed from naming the abstract category to specifying the actual detailed nature of the processes involved. We may, for instance, feel fairly certain that a simple shape such as a spiral snail shell has not been generated from the intrinsic nature of its fundamental units, nor in accordance with some other set of instructions, nor from any template, but has in fact been generated by certain conditions. We may even be able to go a step further and describe the working out of those conditions in terms of relative growth rates of various parts of the mantel edge. But we usually then bog down, and cannot say what conditions determine the growth rates at the various points. However, the mere attempt to determine to what category of form generation a given structure belongs, may suggest revealing lines of experimental attack. In order to try to decide whether a flagellum or a mitochondrion is a unit-generated or a template-generated form, we should have to tackle such questions as to whether these structures can be reconstituted after being broken up mechanically, or whether they can appear *de novo* in systems in which it is implausible to suppose that any template for them is present.

4. Morphogenesis in
Single Cells

*I*N the next three chapters I wish to consider certain examples of the appearance of structural organization in biological systems. It will obviously not be possible even to mention all the innumerable phenomena of this kind which have been investigated. I shall pick out for treatment just a few examples, particularly of cases in which new points of view have been developed or important new facts discovered.

This chapter will deal with some instances in which morphogenesis occurs within single cells, or at any rate within masses of living material of about the same size as a single cell.

In the past, most of the considerations on structural order within single cells have been concerned either with eggs or with protozoa. Many eggs are known which contain various specifically distinct oöplasms which are arranged in a definite morphological pattern. Some classic examples are the Ascaris egg, the series of oöplasms described in Ascidians by Conklin, and the gray crescent in amphibian eggs. The phenomenon of intracellular morphological structure within eggs played an essential role in developmental theory, in which the main concept has been that cleavage nuclei moving into the various different plasms become activated by their surroundings to express certain out of their range of genetic potentialities. Recent developments suggest that the time is now approaching when we shall have to consider whether this basic theory does not need some rather far-reaching elaboration.

Eggs are large cells, and their internal structural organization could be easily verified with the light microscope. The electron microscope, which is revealing so much that was previously invisible, is bringing home to us the fact that intracellular structural organization is a widespread

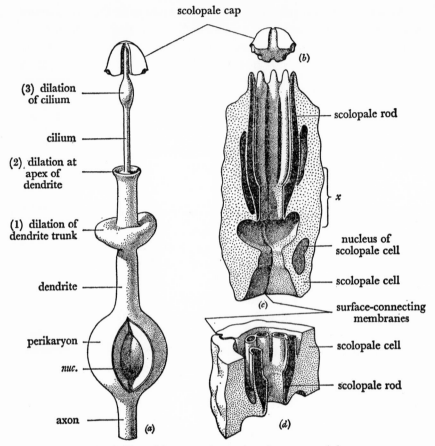

Figure 29. Cell architecture in scolopale organ of locust ear

Figure (*a*) shows the cell body and dendrite of the sensory neuron. This fits inside the scolopale cell shown in (*c*) (the portion marked *x* is illustrated in another way in (*d*)). (From Gray, 1960)

phenomenon and is not by any means confined to eggs. Many types of cells in different organs of animals from all parts of the living world exhibit a morphology which is both definite and quite complex. An example which illustrates particularly the elaboration of the general form of the cell is provided by the scolopale of the locust acoustic organ (Gray, 1960; Figure 29). Many other examples could be found, and I should like to consider in rather more detail a type of cell which I have studied myself, the retinula of the Drosophila eye.

As was mentioned previously, each ommatidium of the eye contains seven main retinula cells and a small eighth cell. Each of these has a rather definite over-all shape, and each contains a number of different constituent parts which are arranged in a definite geometrical relation to one another. One can distinguish at least six such recognizable constituent parts: the nucleus, the rhabdomeres, the rhabdomere tips, the rhabdomere vesicular spheroids, the rhabdomere pigment granules, and

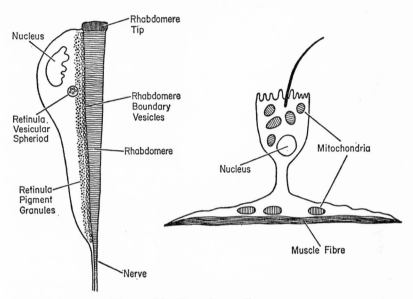

Figure 30. Complex cell structures

The left diagram is the retinula cell of Drosophila, showing the different cytoplasmic components (original). The right diagram is an epithelio-muscular cell from the mesentery of the sea-anemone Metridium. (From Grimstone, Horne, Pantin, and Robson, 1958)

the substance of the optic nerve. Each of these different organelles has a characteristic and distinct appearance in electron microscope sections, and presumably each of them has its own specific chemical composition. In the fully differentiated cell, they are located in particular regions, so that the cell as a whole has a rather definite and precise morphological organization (Figure 30). The same point can certainly be made about many other cells; for instance, in the Drosophila eye itself the structure of the cone cells and of the cells composing the hair-nerve groups are only slightly less complicated and precise than that of the retinulae.

Intracellular organization of this kind raises two main questions: how is the structure brought into being and maintained; and what are the relations between the localized products of differentiation and the genetic factors in the nucleus? As was pointed out above, the conventional classic theory deals with oöplasms, which can be supposed to become separated from one another by plasma membranes as cleavage continues, so that we finish up with a series of cleavage nuclei, each surrounded by a homogeneous cytoplasm, which, however, would be different in different parts of the whole mass of cells. In a retinula, however, we have one nucleus within a cell which contains at least five different cytoplasmic entities, the rhabdomeres, rhabdomere tips, vesicular spheroids, pigment, and nerves. Each of these products must be under genetic control. In fact, we already know examples of genes which cause alteration in one or other of these components without, as far as we know, affecting them all simultaneously. An obvious example is *white eye,* which removes the rhabdomere pigment without producing any noticeable alteration in the rhabdomeres themselves. Now the basic epigenetic theory supposes not only that the genes control the specificity of synthetic processes going on in the cytoplasm but that there is a feedback system from the cytoplasm onto the genes which determines the degree of activity the various genes will exhibit. How are we to envisage this double flow of information in cells in which it is no longer reasonable to think of the cytoplasm as homogeneous?

It is perhaps well to remember some other cases in which the same problem is posed. In more highly evolved organisms, certain types of differentiation tend to be rather strictly alternative. The cells develop, for instance, either into muscle or into epithelium, but not into both. In the more primitive groups, such as coelenterates, we can find cells in which these two types of differentiation are simultaneously present in a single cell body which is provided with only one nucleus (Grimstone, Horne, Pantin, and Robson, 1958; Figure 30). Such phenomena are rather strong evidence against any theory which attributes differentiation to processes of genetic change akin to mutation, which inactivate the genes controlling all types of differentiation except the particular one which the cell will follow. Such theories suggest that in a mammal a cell becomes a muscle cell because all genes involved in other types of process have been inactivated. In the coelenterates, we are confronted with cells which can produce well-developed muscle fibers and, at the same time, carry out other types of differentiation, such as that required to form epithelial structures, such as cilia. The genes controlling the synthesis of muscle proteins, therefore, operate without the inactivation of those which control the formation of these other materials.

Still more cogent evidence against a mutational theory of differentiation would be provided if we could find cases of double differentiation occurring in cells of an animal species in which differentiation is normally alternative. A preliminary report of such a phenomenon has recently been made by Wilde (1960, 1961). He claims that a single amphibian cell may exhibit two usually alternative types of differentiation, not simultaneously but successively in time. Cells of the early neural crest were allowed to develop for some days in culture until they were recognizable as early pigment cells with a characteristic stellate shape and a content of pigment granules. At that time, the culture medium, which had been to some extent conditioned by the cells growing in it, was removed; the cells were washed and placed in a different medium. Wilde claims to have observed in a few cases that a cell which had already become a recognizable pigment cell could, without undergoing division, proceed to develop muscle fibrils, which exhibited a well recognizable cross striation. This report has not yet been confirmed, but we certainly know nothing at present which would make us consider the suggested phenomenon as impossible. We need, therefore, to provide, in our theories of epigenetic processes, for a considerable degree of flexibility. Neither irreversible gene mutations, nor the production of anything like plasmagenes with long-lasting genetic continuity of character, seem appropriate to deal with the facts.

Granted that we have to think in terms of a rather flexible system of two-way information transfer between genes and cytoplasm, the original problem posed by structurally complex cells still remains: Is there or is there not any structural organization of the flow of messages? To put it in another way: Are we to suppose that all the different cytoplasmic regions have an influence on the nucleus, which then puts out the appropriate templates equally in all directions, the structure of the completed cell arising entirely by interaction between the phenotypic products in the cytoplasm; or alternatively, can we suppose that there is any more directed traffic between cytoplasmic region and nucleus?

The possibility that the situation may be of the latter kind is suggested by the fact that, at least in some cells of this type, the nucleus is not symmetrically arranged within the cell body. In retinulae, for instance, it lies at a particular place in the more distal region of the cell. Moreover, the part of the nuclear surface lying toward the rhabdomere tends to be strongly lobed while its opposite surface has a smoother contour. Again, the cone cells which lie just distal to the retinulae contain a characteristic cytoplasmic multivesicular granule. This lies close up against the nuclear envelope; indeed it sometimes appears as though joined to the envelope by thin strands of electron-dense material—and it occupies a definite position in the cell just proximal to the nucleus nearer to the mid-line of the ommatidium (Waddington and Perry, 1960). The whole arrangement strongly suggests a close connection between a cytoplasmic entity and the nucleus; and a connection, moreover, which is structurally limited to a particular part of the cell body. The connection between ergastoplasm and the nuclear envelope, which was referred to in Chapter 2, is again an example of a close relation between cytoplasm and nucleus; and once more, in certain cells at least, there is some degree of definiteness as to the region in the cell where such connections occur. In retinula cells, it is true, connections of ergastoplasm to the nuclear envelope occur both on the outer and the inner faces of the nucleus, though they seem somewhat more common on the inner lobed surface facing the rhabdomeres. In the cells of the amphibian notochordal sheath, which are somewhat flattened in shape, the ergastoplasm tends to come off from the nuclear envelope in the plane of the greatest dimensions of the cell (Figure 13). In some highly specialized cells, for instance, in the protozoan Stentor, an elaborate macronucleus is developed and the different lobes of this have somewhat different properties and seem to be in particularly close metabolic relation with nearby regions of the cell body, particularly the cortex (Tartar, 1960a). It, therefore, seems by no means impossible that the information traffic between nucleus and cytoplasm may be carried out along defined paths rather than operating as though the cell were no more than a homogeneous solution.

We have, as yet, rather little idea of the processes which might bring into being the channels for a directed information transfer of this kind.

This problem is, however, clearly related to the general question of how intracellular structural organization is produced, and to this we must now turn.

There are a number of factors which are usually considered as playing a part in producing and maintaining the structure of individual cells. The most obvious and perhaps the least interesting is gravity. In most eggs there is a polar structure, expressed in the distribution of yolk, which is usually present in the form of granules of greater density than the general cytoplasm and thus becomes accumulated at the lower or vegetative pole. The yolk gradient often plays a very important part in the early stages of development. Experiments in which frogs' eggs are held upside down, so that the heavy yolk falls to the other end of the egg, have shown conclusively not only the epigenetic importance of the yolk gradient but the influence of gravity in producing it (Pasteels, 1951). However, in general, gravity seems likely to be only a relatively minor influence in producing the intracellular structure. In cells such as the Drosophila retinulae the arrangement of the internal contents has no relation to the position of the cells in the gravitational field, and we have to look for something else to explain their structural organization.

Another factor commonly appealed to is the cell cortex. The centrifugation of eggs which are held in definite positions in relation to the centrifugal force usually causes a redistribution of their internal contents, but it is frequently found that if this does not go too far, the development may still be reasonably normal. It can be observed in many such cases that the stratified internal contents become rearranged in their original order. It is usually held that the most convincing explanation of such phenomena is the supposition that the centrifugation has failed to cause any translocation of materials within the tough cortex and that the persisting cortical field has guided the redistribution of the internal materials. Similarly, centrifugation experiments in which the egg nucleus is displaced, either before the maturation divisions, or before the early cleavages, often results in cleavage patterns which show an influence of the cortex as well as of the position to which the nucleus has been displaced. The earlier literature on experiments of this kind will be found in Schleip (1929) and Waddington (1956). In recent years Raven (1958) has laid particular stress on the importance of the egg cortex, which he describes as carrying the "blueprint information" necessary to give structural form and organization to the materials which will be produced in accordance with the chemical specifications implicit in the information carried by the genes.

An overriding importance has also been attributed to the cortex in producing and maintaining the structural organization of protozoan cells. In these, the differentiation of the cortex in the various regions of the organism can usually be made visible, since the cortex incorporates an elaborate structure of granules and connecting fibrils, the so-called infraciliature or kinetics. In most metazoan cells, however, such as eggs of sea urchins or amphibia, a regionally differentiated structure of the cortex is not easy to demonstrate with either the light or electron microscope.

The reality of the cortical field postulated from experiments of centrifugation or reversal of the gravitational field has, however, recently been demonstrated in a direct manner. Curtis (1960b) has found that, if uncleaved eggs of the toad *Xenopus laevis* are treated with solutions containing versene, it is possible by using normal micro-surgical instruments to strip off small areas of cortex from one egg and graft them in defined orientations into a second recipient egg. The regions of cortex grafted in this way are somewhat thicker than the plasma membrane itself. They are, in fact, about $0.5 - 3 \mu$ in thickness, and contain a hyaline layer with a few mitochondria, pigment granules, etc. below the actual plasma membrane. Curtis was able to show that fragments of the cortical layer transplanted in this way carry with them specific morphogenetic capacities. For instance, pieces taken from the gray crescent region will proceed to invaginate wherever they may be placed in the egg, and, moreover, will carry out these movements in the direction corresponding to their own structure rather than that of their surroundings.

The actual form in which this structural information is incorporated into the cortex remains obscure. It is most easy to conceive of it as taking the form of the orientation of elongated molecules within the surface, but there are few, if any, cases in which this can be definitely demonstrated. In the neighborhood of the blastopore in amphibian eggs, some of the cells become transformed into a flask-like shape in which a rotund cell body is connected by a long thin neck to the outer surface. The neck region of such cells exhibits briefringence (Waddington, 1940a), and since this region consists largely of cortex, the birefringence may be an indication of an orientation of the molecules in the cortical material. Apart from such special situations, however, the cortex of most cells is too thin for it to be reasonable to hope to detect molecular orientation within it with the polarizing microscope.

Although there are many phenomena of intracellular organization which could find their explanation in the existence of a cortical field, there are many others for which such an explanation is less convincing. Even if we suppose that different regions of the cortex of a cell could attract to their neighborhood particular types of internal constituent, there are still cases in which we find within the bulk of the cytoplasm an organization which is difficult to attribute to the cortex. One category of cases of this type are those in which the structures involved are considerably smaller than the cell as a whole. This applies, for instance, to the formation of such structures as flagellae, where the arrangement of the nine peripheral and two internal fibrils can hardly be attributed to a similar pattern of growing points on the cortex. Perhaps stronger evidence for the point at issue is offered by instances in which a well-marked internal structure arises within syncytial masses, in which it is very difficult to imagine the presence of anything corresponding to a cortical field. It is worth considering one such example in somewhat greater detail, since it brings us starkly face to face with the essential problem of biological structures for the explanation of which the cell theory is of little assistance to us.

The example to be discussed comes from a somewhat obscure part of the animal kingdom. The Echinoderms belonging to the class Holothuria (sea cucumbers) form in the connective tissue just below the skin a set of small calcareous ossicles. (I shall follow in the main, the account given by Kühn [1955], where reference to the original work will be found.) In certain genera, for example, Leptosynapta, each ossicle consists of two parts which have a form referred to as an anchor and plate (Figure 31). The anchor consists of a long rod which bears at one end a small curved crossbar, referred to as a "handgrip," while at the other is a longer curved member, which gives the whole thing much the shape of an old-fashioned ship's anchor. During development the anchor usually appears first. The plate, when it begins to be formed, takes its origin from a short bar of calcite which branches at each end and then branches again to build up a network of rods. These rods are at angles of 120° to one another, that is to say, they are arranged according to the normal hexagonal crystalline framework of calcite. Their branching is rather regular, and eventually they produce an oval plate with one large hole surrounded by six others in a regular pattern. At the posterior end of the plate, where the anchor lies against it, are a number of smaller holes and above this a raised bow on which the base of the anchor rests.

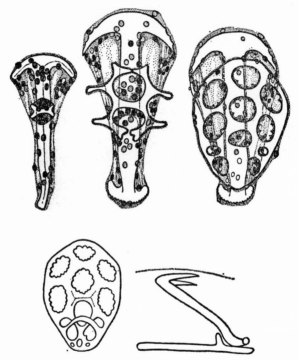

Figure 31. Anchor and plate in Leptosynapta

The upper drawings show three stages of the development of the structures in the syncytium. At the lower left is the plate, and at lower right is a side view showing the relation of anchor to plate. (After Kühn, 1955)

Now the whole of this complicated yet definite structure arises within a syncytial region in which the membranes between the individual cells have disappeared. The syncytia occur in different sizes, and according as they are large or small, produce anchors and plates of corresponding magnitude. It seems almost impossible, in such a situation, to believe that the morphogenesis is to be attributed to the action of a cortex or anything similar. Indeed, it is clear that one of the important factors is the actual crystalline structure of the calcite itself which controls the angle of branching of the rods from which the plate is built. But more subtle and obscure influences arising within the cytoplasm of the syncytium are also involved. For instance, study of the development of abnormal ossicles shows that the plate is largely dependent on the anchor. Usually the initial rod from which the plate develops lies at right angles to the main stem of the anchor, but in some cases this rod may lie parallel

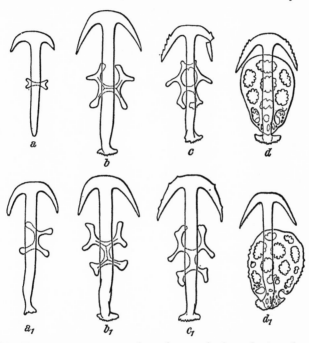

Figure 32. Two arrangements of anchor and plate during development
Diagrams *a*, *b*, *c*, *d* show the usual arrangement during development, with the primary rod of the plate lying perpendicular to the shaft of the anchor. Lower diagrams are the developmental sequence when the rod lies parallel to the shaft; note the different pattern of holes in the resulting plate. (From Kühn after Becker, 1955)

to the anchor stem, and when this occurs the plate develops so that its over-all orientation is relatively normal, though the pattern of holes within it becomes altered (Figure 32). Again, cases can be found in which the anchor has formed in an abnormal way, for instance, by being branched at one or the other end; and in such groups the plates are found to have more or less corresponding abnormalities (Figure 34).

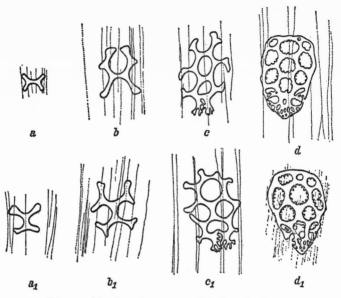

Figure 33. Development of isolated plates

In top diagram with primary rod perpendicular to the folds of the connective tissue; in lower diagram with parallel orientation. (From Kühn after Becker, 1955)

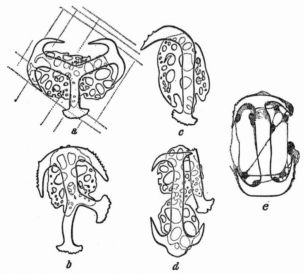

Figure 34. Malformations of anchor and consequential rearrangements of plate in several species of Holothurians (From Kühn after Becker and Wilhelmi, 1955)

However, although these cases make it clear that the anchor can have an influence on the development of the plate, relatively normal plates have developed in a syncytium in which no anchor has appeared (Figure 33). In such cases, the pattern of holes in the plate depends on whether the original rod lies parallel or transverse to the system of transverse folds which is characteristic of the connective tissue.

The formation of these Holothurian ossicles provides a good example of the type of situation which seems to demand an explanation in terms of some internal structure of the cytoplasm. It appears that we are called upon to find some factor which can pervade the whole mass of the syncytium with a basis for the structural arrangement of the calcite laid down within it. Moreover, the factor would have to be of such a kind that could account for the way in which one structure, such as the anchor, may influence another, such as the plate. Phenomena of this kind have led many authors to consider the possibility of an intracellular crystalline or paracrystalline organization of the cytoplasm into what has been spoken of as a "cytoskeleton." For instance, Needham (1936) attached a great deal of importance to this possibility in his discussion of morphogenetic organization in his book *Order and Life* (1936). A more recent discussion will be found in Picken (1960), while the older literature is summarized in Schmidt (1937).

The difficulty has been to detect any such organized framework in the body of the cytoplasm. In order to provide an explanation of the morphogenetic phenomena, the framework would certainly have to have polar properties, and it is most easily envisaged as consisting of the orientation into parallel or other orderly arrays of elongated macromolecular fibrils. One cytoplasmic body in which such paracrystalline organization is well established is, of course, the mitotic spindle; but in the cytoplasm of most nondividing cells, little evidence for the existence of an orientated cytoplasmic framework can be discovered. For instance, if such a framework were well enough developed to have an important morphogenetic influence on the behavior of the cell as a whole, one might expect it to control the orientation of elongated cell inclusions. This is found in certain cells which are either very elongated themselves, such as nerves, or which contain very elongated materials, such as muscle fibrils; in both cases other cell inclusions, in particular mitochondria, tend also to be oriented. There is, indeed, evidence for cytoplasmic orientation in several well-developed columnar epithelia. But what we really require is a demonstration of such cytoplasmic structure in cells which are at the beginning, rather than at the end, of a morphogenetic process; and this has not been so easy to obtain.

One material in which such effects can be looked for is the group of flask cells attached to the blastopore in the amphibian gastrula. These cells have a highly polarized structure and during the process of gastrulation are certainly exerting a pull along their main length. They contain a large number of oval yolk granules. If one of these yolk granules by chance becomes included within a mitotic spindle—an event which happens rather rarely—its long axis becomes orientated parallel to the spindle fibers. On the other hand, within the nondividing cells, no orientation of the yolk granules can be detected (Waddington, 1942a). This seems rather strong negative evidence against the existence of a cytoplasmic framework within these cells. Another early investigation of the orientation of the cytoplasmic macromolecules was the attempt by Harrison, Rudall, and Astbury (1940) to obtain an X-ray defraction pattern from amphibian tissues taken from the limb area at a time when they were known to exhibit strong polarization in their morphological behavior. Again, no signs of such an orientation could be found. A more recent study, equally negative from the point of view of detecting cytoplasmic organization, was that of Crick and Hughes (1950) on the movements of a minute particle of iron within the cytoplasm of a fibroblast placed within a magnetic field. Finally, one may mention the fact that the electron microscope has revealed no clear sign of general pervasive cytoplasmic paracrystalline structures, but one must remember that even the present fixation methods make it very difficult to observe with this instrument the fibrillar organization of the spindle, which certainly exists.

In spite of the predominantly negative nature of this evidence, it would probably be unsafe to dismiss entirely the possibility that orientated cytoplasmic frameworks may be important in certain instances. As an example in which a cytoskeleton has been postulated with some plausibility, it is worth considering the morphogenesis of Micrasterias, which has been rather thoroughly studied by Waris (1950, 1951) and Kallio (1960). This organism is a Desmid. It has the basic structure of the group, that is to say, it consists of two half cells joined by a narrow isthmus in which the nucleus is located. In Micrasterias, the half cells have the over-all shape of flat semicircles, so that the whole organism forms a roughly circular shape and can be considered, at a first approximation, as a two-dimensional form. The semicircle which makes up each half cell is actually cut up into a number of lobes, namely, a central polar lobe and two lateral lobes, each of which is again divided into two, or sometimes more, sublobes. At the time of cell division, the nucleus becomes divided in the neighborhood of the isthmus and the two half cells

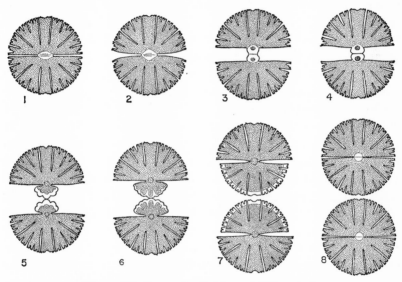

Figure 35. Stages in division of Micrasterias

then move away from one another while each of them forms a new mirror image of itself (Figure 35).

The experimental results obtained with the organism suggest that it has an effective set of three cytoplasmic axes, but that the realization of the potentialities of these axes depends on an interaction between the cytoplasm and the nucleus. The existence of the axes is perhaps most clearly shown by the fact that when, through some abnormality, one of the lateral ones disappears, the resulting half cell continues to reproduce itself without restoring the missing lobe (Figure 37). A clone of such uniradiate cells has been propagated for many years, but since no crossing is possible, the hereditary basis is unknown.

The interaction between the cytoplasmic axes and the nucleus can be demonstrated in several ways. For instance, by centrifuging at the time of division both the daughter nuclei can be forced into one of the half cells, leaving the other half cell anucleate. The two nuclei in the one half cell unite, giving an organism which is presumably diploid (on the supposition that the normal cell is haploid). This diploid cell forms a new half cell whose structure is approximately normal, except that the width of the lobes is increased so that they cannot all lie in the same plane but tend to curl up at the edges. There are interesting quantitative relation-

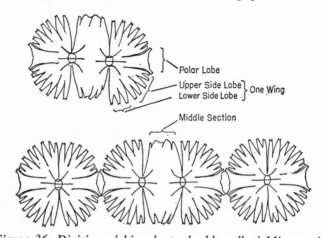

Figure 36. Division of binucleate double cell of Micrasterias

The three lines through the position of the nucleus indicate the plasmastructural units (or cytoskeleton) which control the formation of the polar and two lateral lobes. (From Kallio, 1959)

ships between the amount of nuclear material and the degree of development of the lobes. These may be seen in cases in which, following centrifugation, the cleavage plane at division cuts through the nucleus, giving one cell which contains somewhat more than the normal haploid nucleus while the other has somewhat less; the development of the lobes is proportional to the amount of nuclear material. Perhaps even more demonstrative is the behavior that can be observed in abnormal double cells. If at division the cleavage plane is incompletely formed, the two new daughter half cells may remain attached to one another to give a group of four half cells, the two central ones of which are fused together. If this mishap occurs in normal circumstances, the group of four cells is provided with two nuclei but following centrifugation and other experimental treatments, cases have been found of such double cells provided with only one nucleus at one of the two isthmuses. We can thus have binucleate double cells or uninucleate double cells. The normal division in a binucleate double cell is shown in Figure 36. When a uninucleate double cell divides, the division takes place at both of the

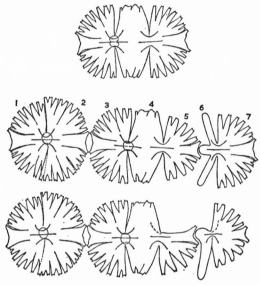

Figure 37. Development of asymmetrical half-cells

The upper drawing shows a double cell with nucleus in the left isthmus. The result of division is shown in the middle row. Half-cell 5, some distance removed from a nucleus, is not as well-developed as half-cell 3; and 3 is worse developed than 2, presumably because it has shared its nuclear supplies with 5. The anucleate half-cell 7 has produced a three-pronged half-cell 6.

The lower row shows a later division. There was a failure of one of the cytoskeletal elements determining a lateral lobe at the anucleate isthmus; thus, the half-cells in positions 5 and 6 both lack this lobe. In later divisions similar asymmetrical half-cells will continue to be produced in this position. (From Kallio, 1959)

isthmuses, although one of these does not contain a nucleus. The situation that arises is shown in Figure 37; and this figure illustrates that the degree of development of the lobes is proportional to the amount of nuclear material and its distance from the developing cytoplasmic regions. It is noteworthy that even the half cell lying farthest to the right is able to develop three quite large outgrowths corresponding to the three cytoplasmic axes, although it has no nucleus of its own and is not even attached to any other cell containing a nucleus. This apparent growth of an anucleate cell is rather remarkable; but Kallio (1959) gives reasons for supposing that it does not actually involve any protein synthesis but is brought about only by the formation of carbohydrates and by swelling due to the imbibition of water. The trilobed form developed by the anucleate half cell seems, however, to give direct expression to the existence of three fundamental cytoplasmic axes.

The situation in Micrasterias is so suggestive of the reality of a cyto-plasmic framework that it seemed worthwhile investigating it with the electron microscope. Only preliminary studies have yet been made, but certain points of interest have already emerged (Plates xiv and xv). In the first place, there is no obvious sign in the cortical layer of the young half cells of structures which could embody the three axes. The half cell soon becomes molded into a trilobed shape, but the thickness of the cortex seems to remain the same all round these lobes and in between them. On the other hand, within the body of the outgrowth, the cyto-plasm has a structure radiating away from the nucleus. The nucleus itself is of a very remarkable character. Quite shortly after division (i.e., in what should correspond to telophase), the nuclear contents are relatively homogeneous except that they include a large number of separate nu-cleolar-like masses. Both the nucleolar material and the general contents of the nucleus stain heavily with uranyl acetate. The most peculiar feature is that the nuclear material seems to be drawn out into filaments extending into the cytoplasm of the newly developing lobes. In the period just after division, no definite nuclear envelope can be seen bound-ing these filaments. Soon after this a double-layered envelope of the normal type begins to appear first in the concave folds of the nuclear surface between the processes which extend out into the cytoplasm (Plate xv). From these places the envelope spreads until it eventually covers the whole nuclear surface; but it lies, not at the boundary of the material which is electron dense in stained preparations, but usually slightly inside this edge, so that even when the nucleus is completely covered with an envelope, it has electron-dense material lying in contact with it. The general appearance suggests that just after division material is being drawn off from the nucleus and spreading into the cytoplasm along roughly radial lines of flow. This type of behavior would fit in very well with the evidence suggesting that the degree of development of the new half cells depends on influences proceeding from the nucleus, as we saw in the case of the uninucleate double cells described above. It is, however, very unusual to find nuclear material becoming dispersed in the cytoplasm in this way, although Moses (1960) has illustrated rather similar appearances during prophase in a crayfish. This suggests, I think, that the nucleus of Micrasterias is not a normal haploid or diploid nucleus, such as is found in organisms with sexual generation, but is something perhaps more comparable to the macronucleus of ciliates. It is worth noting that the number of chromosomes which have been seen in normal cytological preparations of Micrasterias is very large, probably over 200.

In Micrasterias, then, the electron microscope shows in the young half cells a radiating arrangement of fibrous cytoplasm. This suggests that in this form there is really something which has the character of a cytoskeleton. Waris (1950) claimed that if the cells are allowed to divide in fairly strong sucrose solutions, with the light microscope one can see within the new half cells three strands corresponding to the three axes which control the formation of the polar and two lateral lobes. In some of the electron microscope pictures also the nucleus seems to be drawn out predominantly, but not exclusively in three main directions. Altogether then we have in this form rather good evidence of an organized fibrillar structure of the cytoplasm which controls the morphogenesis.

If we refer to such a structure as a cytoskeleton, this name may seem to imply the existence of a more or less static and persisting cytoplasmic framework. A valuable corrective to any such simple idea is provided by observations with the time-lapse movie camera. One of the first people to use this instrument for studying single cells was Dr. F. G. Canti, who in the 1930s made a number of films on tissue cultures set up at the Strangeways Laboratory, Cambridge. Some of the films he took then, with very simple illuminating systems, such as dark ground illumination or an unfocused condenser, have scarcely yet been bettered with all the resources of phase contrast. His films showed not only the formation of pseudopodia and moving frills and filaments on the surface of the cells, which might have been expected, but also quite extensive and rapid movements of particles within the cytoplasm. For instance, most of his films were made on the chick fibroblasts which contain many filamentius mitochondria, and these could be seen moving actively about within the cytoplasm and dividing by transverse fission. Since those pioneering days many time-lapse films have been made on a multitude of different types of cell. Nevertheless, many biologists still seem to experience something of a shock when they see such films and realize that all cells have to be considered as highly active bodies in which movement of the internal constituents is continuous and uninterrupted. Time-lapse films, of course, exaggerate the speed with which these movements are carried out, and accepted at their face value, make an impression which is perhaps exaggerated. On the other hand, the point they bring home so forcefully, that cytoplasm is always in a state of physical activity, is a perfectly valid one. To have it exaggerated in this way is useful to counterbalance our tendency to envisage cells in terms of the static pictures presented by ordinary microscope preparations.

In our laboratory we have recently shot some films of dividing Micrasterias. They show clearly that the internal contents of the Micrasterias cell is continually jiggling about and is not simply staying in place. If we are to think of the cell as possessing a cytoplasmic framework characterized by three main axes, we have to remember that this framework must be something of a dynamic nature, something that is to say, which is not merely brought into being at one time and then persists unchanged, but a system which requires to be continually maintained in being by active processes. The same considerations apply to many, and possibly all, cells. Certainly in all cells of which I have seen time-lapse films, the movement of the internal contents is a noticeable feature. I feel fairly confident that if we had the technical means to observe, over some period of time, the structure of one of the types of cell whose morphological organization we have been considering, such as the retinula cells, we should observe continual slight alterations in position of the nucleus, the pigment granules, the ergastoplasm, the vesicular spheroids, etc. The rhabdomeres themselves are structures with material continuity from one end of the cell to the other, and it is perhaps likely that there is rather little movement within the structure. But the main problem of intracellular morphological organization with which we have been concerned in this chapter is the orderly arrangement of a number of cell parts which are separated from one another by the ground substance of the cytoplasm. The argument I am putting forward now is that these must keep their relative positions under the influence of continuously acting forces rather than because they are attached to any form of rigid framework. Moreover, the action of these forces must be subject to considerable short-term variation in intensity, so that the organelles have quite considerable relative motion.

A dramatic demonstration of the reality of active intracellular forces is provided by Tartar's description (1960b) of the reconstitution of a fragmented unicellular organism, Stentor. If an individual is minced, and then pieces allowed to reaggregate, they sort themselves out, much as disaggregated cells do, so that the normal structure is eventually restored.

It is not entirely easy to see where we should look to find a suitable system of interorganellar ordering forces. What we require is a set of attractions and repulsions between entities of different chemical composition measuring from 1 to 10 microns or so in diameter and separated by distances of similar or somewhat greater order of magnitude. We mentioned in Chapter 3 two main types of relatively long-range attractive forces which are commonly appealed to in connection with the formation of subcellular structures, the London-van der Waals and complementary charge forces acting between relatively extended surfaces. When we are dealing with entities which are separated by a gap filled with an aqueous medium, we have also to take account of repulsive forces due to the formation of an ion atmosphere around each particle. All these forces, however, although long-range in atomic terms, only become important over distances which are very small in comparison with those between the organelles within a cell. The maximum range at which we can expect them to be important is of the order of a few hundred angstroms. It is very difficult, therefore, to believe that they are importantly involved in maintaining the intracellular structural organization of the kind that we are discussing.

Another type of force, which may prove to be more important in this connection, might be generated by the spontaneous alterations in shape of the cellular organelles. These structures are composed of macromolecules, such as proteins, lipids, carbohydrates, combined in various ways. It is well known that a protein molecule may exist in a number of structurally different forms, and the same is probably true of the other macromolecules involved. Alteration from one molecular configuration to another is quite drastic in some protein species and has, of course, been utilized in the development of contractile mechanisms based on actin and myosin. One is probably justified, however, in considering that in principle all the macromolecules in biological materials are capable of some change of molecular configuration. Such changes will probably often occur spontaneously, being brought about by processes of fast fundamental particle transfer, such as proton and electron transport processes which are controlled by quantum laws (see Szent-Gyorgi, 1961; Schmitt, 1960).

The gross physical results of such changes in molecular configuration have not yet been fully studied in detail. It seems rather likely, for instance, that they will provide the explanation for the as yet quite mysterious process of cytoplasmic streaming or cyclosis, which is such a striking feature of many plant cells. They might also give rise to more or less rhythmic pulsations in organelles such as mitochondria, pigment granules, and the membranes of the ergastoplasm. Any such changes in shape or volume, if at all regularly repeated, could generate forces of attraction or repulsion between the structures concerned.

Perhaps the simplest of these forces are those which produce the so called Guyot-Bjerknes effect. This effect, which has been known for over a century, arises essentially from the fact that two pulsating spheres immersed within a liquid will attract one another if their pulsations are exactly in phase, while they will repel one another if they are out of phase by 180°. The forces involved are hydrodynamic ones, and appear as a direct consequence of the fundamental theorem due to Bernoulli, that the pressure within a body of steadily flowing fluid is smaller, the greater the velocity (Figure 38). Fabergé (1942) has considered the application of such forces to explain the pairing between homologous chromosomes in meiotic prophase. He based his discussion on the simplest possible model, namely, of two spherical objects which pulsated with radial symmetry over their whole surface. The requirement that the pulsations must be exactly in phase if a long-lasting and consistent force is to result would, he suggested, be met because of resonance between pulsators of approximately the same frequency. In the real situation within cells, we presumably have to deal with a much more complicated system. There must be many pulsating bodies of different sizes and presumably, at least initially, somewhat different periods. However, these periods would be expected to influence one another not only through resonance phenomena arising from the presence of the medium in which the bodies are immersed, but also because the changes in molecular configuration which give rise to the pulsations will also alter the electric field around the particles. Again, in large macromolecules and still more in particles of larger dimensions, the pulsations are not likely to be regularly symmetrical and there may be several pulsating sites located at specific places on the surface of the organelles. These forces may be effective over distances at least comparable in size to those of the structures by which they were produced, or indeed a few times larger. They, therefore, might well come into question as the main interorganellar ordering forces within the cell. They remain, however, in need of much further study.

Figure 38. Guyot-Bjerknes force

The Guyot-Bjerknes force depends on the fact (Bernoulli's theorem) that pressure in a flowing fluid varies inversely as the square of the velocity. Consider two spheres immersed in a fluid and pulsating in phase. In the drawing on the left, they are represented while both expanding. The expansion of 1 will set up a current flowing past 2 (arrows A and A'), while that of 2 will produce current B and B'. Thus the flow relative to the left side of sphere 2 will be faster (A + B) than that on the right side (A − B), and the pressure will therefore be less, so that 2 will tend to move to the left. When the spheres contract (drawings on right), the relative flow will again be such that sphere 2 moves to the left. Meanwhile, the forces acting on sphere 1 will be the mirror image of those on 2. The net result is that the two spheres attract one another. (After Fabergé, 1942)

Because we have so little understanding of the forces which produce and maintain intracellular organization, it is interesting to examine some of the biological phenomena in which a well-recognized cell type appears in an unusual form. One of the most striking examples of this is provided by the interconvertibility of certain unicellular organisms between an amoeboid and a flagellate organization. It has been known since the end of the last century that organisms of the family Bistadiidae, such as Naegleria species, move around in an amoeboid form when they are on a relatively dry substratum in the presence of bacteria, but when they are surrounded by more or less pure water, they develop flagellae at one end and the whole body becomes elongated into a typical flagellate type. The most recent student of the phenomenon is Willmer (1956, 1958). He showed that the stimulus to the transformation is not a mere lowering of the osmotic pressure, but that the change from amoeboid to flagellate form is favored by low concentrations of certain salts, such as sodium bicarbonate, sodium lactate, and sodium phosphate, while it is inhibited by certain other substances such as lithium salts, magnesium chloride, and the sulphate ion. Willmer has used these facts as the basis for some interesting but perhaps over-ambitious speculations concerning polarized differentiations of cell type within multicellular organs of the metazoan body, but into these I do not propose to follow him here. The main point in our present context is that comparatively simple changes in the external environment can bring about a profound reorganization of the whole intracellular structural organization. Another similar example is the conversion of Trypanosomes into the blood-stream form by the action of urea (Steinert, 1958).

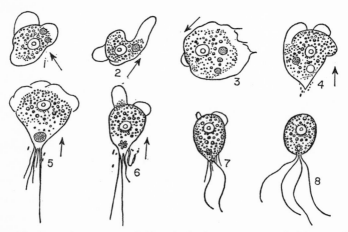

Figure 39. Transformation of Naegleria from an amoeboid to a flagellate form (From Willmer, 1956)

Unfortunately, here again, as in so many instances in the experimental study of development, the fact that we have discovered mechanisms for pulling certain triggers tells us very little about the nature of the mechanism which the trigger sets in action. The changes involved in the transformation of Naegleria are relatively complex (Figure 39). The amoeboid form has a very indefinite shape, and rounded pseudopodia may appear at any point on its surface, while the contractile vacuoles, of which there are one or two in the cytoplasm, seem to appear and disappear again without any definite location. When the cells begin to transform into the flagellate form, the first sign is the appearance of a definite polarity. This first affects the position in which the pseudopodia are pushed out, and within a short time large rounded pseudopodia are found only at one end, which may be considered the anterior end. At the other end of the cell a number of thin threadlike pseudopodia make their appearance at this stage and the contractile vacuole takes up a position in this neighborhood. Shortly afterward one or more (usually four) flagellae appear in the same neighborhood and immediately start to beat. By this time the formation of rounded pseudopodia at the anterior end has been curtailed and it soon ceases completely. The cell is thus transformed into a well-organized flagellate type with strongly marked polarity. What forces have brought the contractile vacuole to its definite position, have fixed the place at which the flagellae grow out, and have maintained the definite polarity of the cells? Unfortunately, as Willmer writes, "almost nothing is known of the nature of this change." It is quite easily reversible: the flagellate forms changing back into amoeboid types either if the medium dries up or indeed after a fairly short period even if they remain submerged in water. The change is, therefore, not to be interpreted in genetic terms, but is solely a rearrangement of the existing material, without any permanent alteration in its character.

Rearrangements of this kind are, of course, well-known in metazoan cells and are well-exhibited when, for instance, cells from a fragment of tissue are allowed to migrate in tissue culture. They have been referred to by Weiss (1939) as "modulations." The amoeboid-flagellate transformation is only a particularly striking example of this general category of cellular transformation, but it brings home to us very forcefully how little we yet know about the actual mechanisms of intracellular structural organization.

Another way of approaching the subject is to study the effect of mutant alleles on cells which have a well-defined morphology. We have begun to look with the electron microscope at some of the numerous abnormal types of ommatidia produced by genes in Drosophila. So far we have only made a preliminary survey of a number of different types, searching for some whose development it seems interesting to follow in detail (Plates XXI to XXIV). In many of the mutant phenotypes the general anatomical relations between the groups of cells making up the ommatidia are considerably disarranged. Some of these will be mentioned again in the next chapter when we are discussing the relations between small numbers of cells. One point which emerges relevant to the present consideration is that we can find, for instance, in *rough* or *polished,* single isolated retinula cells which have developed all their normal intracellular structures including the rhabdomere. Thus, although the rhabdomeres normally begin to form by a folding of the cell membranes in the center of a group of retinula cells, the process does not essentially depend on an interaction between similar retinula cells, but can be carried out by an isolated cell which is surrounded by cells of a different constitution.

In the irregular ommatidia the shape of the rhabdomeres is often quite abnormal. One also gets the strong impression that the volume of rhabdomere material, in proportion to the volume of the cell as a whole, is by no means so constant in these abnormal cells as in the wild-type. Some rhabdomeres seem much too large for the cells that bear them; others too small. Owing to the difficulty of making long enough sets of serial sections for examination in the electron microscope, it is difficult to be quite certain of this point. There is no doubt, however, that the rhabdomere material may be located in abnormal positions in the cell. Some retinulae, for instance, in *split* eyes, carry more than one rhabdomere, and one of these may be on the outer side of the cell, where it abuts against a pigment cell. The rhabdomere pigment granules may also be abnormally located, lying in *polished,* for instance, far away from the rhabdomere. The internal architecture of the rhabdomere itself may also be disarranged, to a greater or lesser extent. In normal eyes the diameter of the individual tubules is remarkably regular, while in some of the mutant types, such as *Glued,* they show considerable variation. Again, in eyes such as those of *rough,* one finds patches of tubules whose orientation is not concordant with that of the neighboring material.

We have so far, however, found no mutant in which there are retinula cells from which rhabdomere material is completely absent (except in obviously degenerating cells, such as may be found in *polished*). It seems indeed very unlikely that any of these genes control directly the synthesis of the material from which the rhabdomere is built up. It is much more probable that all of them act to produce abnormal conditions within the cell as a whole, and that the disturbances we see in the rhabdomeres are the result of the reaction of normal rhabdomere material to these abnormal conditions.

Conclusion

The phenomena of morphogenesis in single cells confront us with two main problems. The first arises from the fact that single cells often exhibit a complex and rather definite architecture. The basic problem of morphological pattern—of why, for instance, a dog has, roughly speaking, a leg at each of its four corners—is not a problem which arises only in masses of cells, and which we can therefore hope to explain in terms of the interactions between cells. It confronts us also when we examine some types of single cell, such as the retinulae of the insect's eye. We shall have to find for it an explanation which is not dependent on the properties of cells as wholes, but which deals with factors which can operate either within individual cells or between cell groups.

When we contemplate various factors which might operate in this way, we are immediately brought up against the second main point which has been discussed in this chapter. The internal constituents of cells are usually, if not always, in movement. Any system of intracellular ordering forces which we might be tempted to invoke must, if it is to be plausible, result in types of structural stability which are dynamic rather than static. As we have seen, the distances involved in intracellular architecture are too great to be explained in terms of the forces usually considered to operate between molecules; but, I argued above, forces of a different and as yet rather unexplored kind may arise between macromolecules or cell organelles which are themselves flexible and contractile. It is still more obvious that ordering forces which act on a supracellular level are even less to be explained in terms of conventional intermolecular interactions; but here we have the possibility that the order arises from reactions between flexible and motile cells. The next chapter will be devoted to a consideration of reactions of this kind. As we shall see, intercellular interactions, particularly those depending on properties of cell membranes, suggest ways in which some examples of biological structure can be accounted for. But there are other examples for which such mechanisms do not seem adequate, and where the facts suggest the operation of pervasive forces which operate throughout a whole mass of living material independently of the fact that the mass is cut up into individual cells.

5. Multicellular Morphogenesis

In Small Groups of Cells

THE organization within a single cell must result rather directly from interactions between the various gene-action systems. We need now to proceed to consider more complex types of order. The first of these is morphogenetic organization in which the units are individual cells. The characters and properties of these cells will, of course, ultimately be determined by their genetic content and the particular gene-action systems which are in active operation within them. The interrelations between the cells, however, give rise to phenomena which, if they could be fully analyzed down to the gene-action systems, would prove to be of another order of complexity. We cannot actually carry out such analyses, but must be content to push our understanding of the morphogenetic events down to an intermediate level of analysis. The forms and structures produced by small numbers of cells often have a considerable simplicity; and the first stage of analysis of their method of generation leads us, not right back to the gene-action systems but to a consideration of one particular cell organelle, namely, the plasma membrane. It is this external layer of the cell which controls the processes of cell movement, cell contact, and cell adhesion, by which the structural organization of groups of cells are brought about.

A recent major field of experimentation has been that of the reaggregation of cells. This has depended, in the first place, on the development of methods of disaggregating tissue, that is to say, breaking down the primary adhesions between cells in tissues. Within the suspensions of single cells so obtained, various processes of reaggregation can be seen to proceed. The earliest work of this kind dates back many years and was carried out with sponges. More recently, there have been an important series of observations on amphibian cells by Holtfreter (1947), Townes and Holtfreter (1955), and Curtis (1961); and on avian and mammalian cells by Weiss, Andres, Moscona and others (a recent review is Moscona, 1960). The subject is closely allied to the study of the movements of cells which lose contact spontaneously in tissue cultures; a subject which has been studied for many years particularly by Weiss (cf. 1958), and more recently by Abercrombie (1961).

When suspensions of cells of different types are mixed and then allowed to settle on a surface, the sequence of events is usually as follows: After moving around as individuals for some time, the cells carry out a primary reaggregation into clumps, which at first consist of relatively few cells, but then the separate clumps fuse into larger masses. The next step is for the cells within a mass to sort themselves out, so that they form regions consisting of cells of similar type. These regions may become arranged in some fairly regular order, for instance, with one type of cell on the outside of the mass, another type on the inside, and so on. Finally, the whole mass of cells may undergo morphogenesis and differentiation. This is sometimes most surprisingly complete. When disaggregated cells are placed in some position which maintains their life for considerable periods, they may, after reaggregation, eventually develop into extremely complex and precisely formed structures. Weiss (1950); Weiss and Andres (1952), have described some very striking instances of this when disaggregated cells have been placed in the neutral environment of the mesenchyme filling the tail fin in amphibian tadpoles, and also when they have been introduced into the blood vessels of the chorioallantois of the chick embryo. These beautifully formed final organs are, however, the end stages of a long sequence of developmental processes. The basis for the whole later reorganization must be laid in the initial stages in which the cells come together and sort themselves out into regions, and it is these primary stages which present the real challenge to our understanding.

The first reaggregation of cells into clumps seems often, and perhaps always, to be unspecific. In our laboratory, Lucey and Curtis (1959) have recently made films of the reaggregation of early amphibian embryonic cells (Figure 40). In single cell suspensions, the cells move about at random on the surface where they rest and send out randomly directed pseudopodia. When these random movements bring them into contact, they sometimes, but not always, tend to stick together and to move over each other's surface. It can be observed that cells of different type are usually as ready to maintain contact in this way as cells of similar type. The initial small groups of cells formed within a mixed suspension are, therefore, mixed in type. This unspecificity persists until the small groups have fused into relatively large masses. The sorting out into regions consistently made up of a single type of cells is a secondary stage.

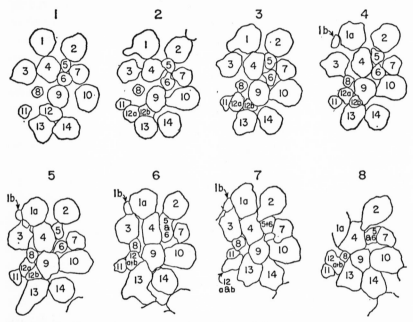

Figure 40. Stages in reaggregation of amphibian blastula cells traced from a time-lapse film (From Lucey and Curtis, 1959a)

When this sorting out occurs, it may take place according to several different rules. What in fact do we mean by "like cells" in this connection? The experimental results show that the effective similarities are different in different cases. If cells of two histological types from one species are mixed with cells of the same two types from another species, they may in some instances become sorted out according to histological type, while in other instances, they become sorted out according to species. For example, if chick and mouse kidney cells are mixed with chick and mouse chondrogenic cells, they become sorted out into groups of kidney cells and groups of chondrogenic cells, but each of these contain mixtures of the two species (Moscona, 1960). On the other hand, Curtis (1961) found that when mixtures of histological and species types are made within newts, sorting out occurs according to species rather than according to histology.

The mechanism of this sorting is one of the most interesting questions at present under investigation. In general, it is clear that the cells are continually moving over and around one another within the mass, and the groups into which they eventually become sorted out are those in which the cells adhere most closely to one another. One of the major questions is whether we can regard this sticking together of like cells as due to the existence of cell-specific adhesion or attraction. This would certainly seem to provide the simplest and most immediate explanation of the phenomena, and several suggestions have been made as to how such cell-specific adhesion might be supposed to arise.

It has been suggested, for instance (Weiss, 1958), that the surfaces of some cells may be related to one another as chemical complementaries, as antigen is supposed to be related to antibody (the "lock and key" hypothesis). It is, of course, very easy to draw idealized diagrams indicating such a possibility, but it is rather more difficult to establish it as actually existing; and there are in fact some quite considerable difficulties in the way of such a theory. Antibodies to cells may, indeed, destroy their reaggregating properties (Spiegel, 1954); but this does not demonstrate that antibody-antigen reactions are actually involved in the forces holding the cells together, since the antibodies may be having a more general effect on the whole cell constitution and metabolism, and their influence on the reaggregating properties may be secondary to this. Perhaps the major difficulty to accepting that chemical complementariness is the basis for cell adhesion arises from the consideration of spatial magnitudes. The forces involved in chemical complementary attractions are short-range forces, operating over distances of only a few angstroms. However, the electron microscopical evidence suggests that cells very rarely approach one another as closely as this, admittedly, there is always the possibility that the separation between cells, as seen in electron microscope preparations, may have been somewhat enlarged by the processes of fixation, etc. We can also consider whether it is reasonable to expect them to get so close. In order for them to do so, the fluid medium between the cells would have to be drained away between the surfaces which are approaching one another; and when one is getting down to separations only a few angstroms across, viscosity becomes of enormous importance. Curtis (1961, 1962), who has recently reviewed the matter, comes to the conclusion that it is extremely improbable that any forces of attraction could be generated which would bring the cells into such close contact that chemical complementariness could operate between them except over very small areas.

Another suggestion that has been made (e.g., Steinberg, 1958) is that cell specific attractions arise from the presence of complementary regions distributed in corresponding patterns over the whole cell surface. In this case, the distance between the elements in the corresponding patterns is supposed to be quite large, that is to be measured in terms of several microns or tens of microns. The suggestion has, perhaps, a certain validity in so far as cells have in some cases specific regions in which they become attached to other cells by bodies, such as desmosomes. We shall consider these at a slightly later stage in this discussion. It does not, however, appear very likely that the presence of specific attracting regions of this order of magnitude could account for the effective sorting out of cells. The two patterns would only complement one another when arranged in the correct orientation. If, for instance, the pattern on each cell took the form of a series of parallel lines, the patterns of two cells would only complement when the lines of one cell were parallel to those of the other. Further, when we are dealing with long-range forces, we have to take into account the interference between two attracting points of the presence of another nearby point which exerts a repelling force. The specificity of attraction exerted by a definite pattern of regions, therefore, extends only a comparatively short distance from the surface, and becomes blurred and ineffective at distances much greater than those between the separate elements in the pattern.

Attempts to provide a mechanism which would bring about cell specific attractions have run into some difficulty for these reasons. It is, however, by no means certain that the mere fact that cells become sorted out by type necessarily calls for the postulation of cell specific attractive forces. Curtis (1961) made the important observation that in mixtures of amphibian embryonic cells the specific sorting can be altered by changing the time relations. If a mid-gastrula stage of an amphibian egg is disaggregated and the cells then allowed to reaggregate, they normally form masses in which the ectoderm cells make up the outer sheet, while the endoderm cells form the center of the mass and the mesoderm cells lie roughly between the two. If, however, one takes a culture of endoderm cells and allows them to reaggregate for a number of hours before adding

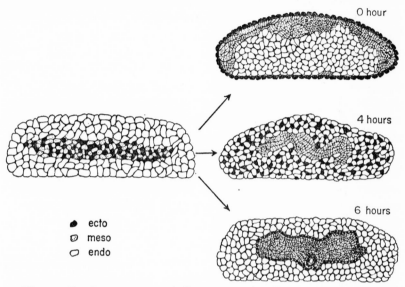

Figure 41. Arrangement of disaggregated cells after sorting out

Disaggregated ectoderm and mesoderm cells from mid-gastrulae of the toad *Xenopus laevis* were wrapped round with endoderm cells, after the latter had been allowed to reaggregate for 0, 4, or 6 hours. The drawings on the right show the sorting out which had occurred 16 hours later. If the endoderm had not started to reaggregate before the addition of the other types, the ectoderm would eventually cover the upper surface of the mass with mesoderm below and endoderm at the base. After 4 hours preliminary reaggregation of the endoderm, the meso-derm finishes in the center with a covering of mixed ectoderm and endoderm. After 6 hours reaggregation of endoderm, the final order is endoderm outside, mesoderm in the center, and ectoderm in an intermediate layer. (From Curtis, 1961)

the ectoderm and mesoderm cells, the results are quite different. The character of the sorting out and arrangement depend on length of time for which the endoderm is allowed to reaggregate on its own (Figure 41). There is other evidence that the adhesiveness of developing cells is chang-ing as time passes. Mookerjee, Deuchar, and Waddington (1953) showed this by applying a standard disaggregation procedure to consecutive stages of the developing notochord in amphibia, and Curtis (1957) has described similar results with the gastrula stages. Thus, it seems quite possible that the sorting out of reaggregating embryonic cells is a consequence of the different cells passing at different rates along a series of changes in generalized adhesiveness. A mixture of cell types which differed in general adhesiveness would, of course, be expected eventually to become sorted out into regions of cells of similar type. It may well be that no cell specific adhesiveness is involved at all.

Another factor in the situation to which attention has recently been drawn, is the extracellular matrix, that is the material which is normally secreted from tissue cells and lies in the interspaces between them. This material, which often contains RNA (Curtis, 1958), may, when present in large quantities, exhibit signs of a fibrilla ultrastructure, and there is every reason to believe that this has great importance for the morphogenesis of loosely packed tissues, such as connective tissue, and in directing the growth of nerve fibers, the migration of pigment cells, etc. (Weiss, 1950, 1958). Extracellular matrix also plays a role in some intercellular inductive reactions (Grobstein, 1955). Some authors, including Moscona (1960), suggest that it may have a very general influence, including that of being an agent of specificity. Moscona suggested that we should "examine the extracellular matrix as a cell-integrating system, endowed with specific cell-directing activities that affect movements and associations of cells; thus, a system combining the functions of a cell-binding framework and an information network." Although there may be something in this notion in connection with some of the loosely dispersed tissues of the adult, it is difficult to see how it can have much relevance to the closely packed aggregations of cells, such as we commonly find in early embryonic organs. Moreover, it is worth bearing in mind the remark of Grobstein, who uses the extracellular matrix to explain many of his own results on induction, but also warns us: "As with so many notions founded largely on ignorance, subsidiary *a posteriori* assumptions can be made on the matrix model to explain nearly anything."

It is perhaps more profitable for the immediate future to try to consider the application of general physical theory to the problem of the forces which are likely to act between cells. When two cells approach one another we are dealing with the attractive and repulsive forces which can operate between rather extended areas. In effect this means the attractive long-range London-van der Waals forces, counteracted by repulsive forces arising from the formation of ionized layers from the medium between the surfaces. The London-van der Waals forces are likely to be much the same in all cells and to fall off approximately as the inverse second power of the distance between the two surfaces. The repulsive forces will also fall off exponentially with distance. From a study of the properties of colloidal systems in general, Verwey and Overbeek (1948) have shown that the London-van der Waals and electro-static forces tend to come into balance at two separations, one of which is about 5Å and the other about 100–200Å. The latter spacing is the order of magnitude of the distance usually found with the electron microscope between tissue cells—unless there is a secondary secretion of intercellular matrix between them when, of course, the separation may be much larger. A very close adhesion between cells corresponding to the smaller of the two distances mentioned by Verwey and Overbeek has sometimes been found over small areas of the cell surface.

The electrostatic repulsive forces and the exact balance of forces between approaching cells are considerably affected by a number of different factors. One of these factors is the nature of the substances in the surface which determines the degree of ionization that can occur. Another factor is the nature of the medium, which again has an effect on ionization; in particular this will be affected by the presence of divalent ions, such as calcium, which is well known to play a large part in cell adhesion. And a final important factor is the degree of packing of the surface, that is to say, the area of surface per charged element—the charged element may be a molecule, or part of a macromolecule, such as a protein.

Curtis (1960, 1961) has pointed out that cell movement may be expected to alter the degree of packing in the surface, and thus to affect the electrostatic repulsive forces and the adhesiveness between cells. When a cell moves over a substratum, there must be a point of adhesion between the cell's surface and the solid that it is moving over, so that a movement of the cell body can be brought about in relation to this fixed point. This movement must cause a shearing stress in the cell surface around the fixed point. Now, shearing of a surface tends to alter its intra-surface viscosity. The theory of this process has been discussed by Joly (1956). A change in viscosity will lead to an alteration in area; there is an expansion if viscosity falls, a contraction if it rises. This in turn will bring about an alteration in the surface area per element, that is to say, in surface packing, which as we have seen is one of the factors determining cell adhesiveness. Curtis argues that cell movement will thus give rise to a dynamic adhesiveness, in addition to the original static adhesiveness which we may think of as having been responsible for the original fixed point of contact with the surface over which movement is taking place.

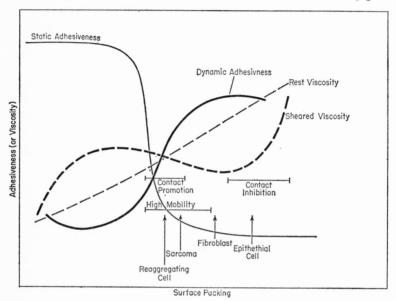

Figure 42. Relation between surface packing and various types of adhesiveness (or viscosity)

The ranges within which various phenomena may be expected, and the approximate positions of several types of cells are indicated. (After Curtis, in press)

The effect of shear on viscosity depends on the initial packing of the surface. Cells which initially differ in this way will show different properties in the degree to which they develop dynamic adhesiveness. This is illustrated in Figure 42. There will be a great difference in behavior between cells whose initial surface packing is such that shear produces a large increase in dynamic adhesiveness and cells on which shearing does not have such a marked effect. The former will show the phenomena which has been described as contact promotion; that is to say, when two such cells start to move over one another, the shearing in their surfaces will lead to a large increase in their adhesiveness, and the cells will tend to stick together and to continue moving over one another. In cells which have initially a higher surface packing, shear will lead to little or no increase of adhesiveness; and these will exhibit contact inhibition, that is to say, they will tend not to move over one another with the result that their random movements will carry them away from each other.

Curtis claims in this way to show how properties adequate to explain the phenomena of sorting out of reaggregates, and many of the mutual influences of moving cells in tissue culture, can be explained without the postulation of any specific intercellular forces of attraction or adhesion. The hypothesis sounds like a very attractive simplification of a complex series of processes and represents a very new approach to this puzzling field. However, the hypothesis undoubtedly requires much further elaboration and experimental testing before it can be considered fully convincing.

Curtis's theory is in terms of cell membranes which are pictured as extensive, more or less homogeneous, sheets. Unfortunately there is little published on the electron microscopical investigation of the surfaces of reaggregating cells. We know that many embryonic cells have surfaces whose microstructure is anything but simple but contains many small projections, foldings, etc. The possibility that such surface processes may play a part should certainly not be overlooked.

It must be remembered also that after cells have been for some time closely associated in tissues, they may develop very definite structures concerned with their adhesion. Similar structures may also be found in the cells of the embryonic stages when primary morphogenesis is going on. Balinsky (1959) has pointed out that in early sea urchin embryos three types of cell contact can be found. First, ordinary contact in which long stretches of two plasma membranes lie close together with a gap of some 140Å between them; this presumably being filled by some electron light intercellular substance. Second, in some cases neighboring cells seem to be held together by the formation of relatively large, but elaborately folded, interdigitating processes. Third, occasionally intercellular connecting bars appear, usually in groups forming structures similar to the desmosomes described in adult tissues. I have recently added a fourth type of intercellular adhesion among embryonic cells. The blastomeres in the cleavage stages of the eggs of the mollusc *Limnea peregra* are connected by small bodies (one or two micra in over-all dimensions) which seem to be formed by a very elaborate knotting together of the plasma membranes of two or more cells (Waddington, Perry, and Okada, 1961a and Plate xvi). In the early embryonic stage of amphibian cells, of the kind studied by Curtis, contact seems usually to be of the first type described by Balinsky, but as we shall see later more localized adhesion bodies are probably beginning to develop, at least from the neurula stage onward if not earlier; and the possibility that such localized adhesions may play a part in cell reaggregation, needs to be borne in mind.

It is now time to consider the organized morphogenesis that may occur within small groups of cells and bring into being structures of well-defined anatomical shape. It has often been suggested that, in some cases at least, the prime agent which causes the molding of a group of cells into a mass with a definite configuration may be forces exerted within or between the cell membranes. In one of the simplest examples of morphogenesis, the drawing-together of the presumptive notochord cells into an elongated cylindrical rod, the whole process could be explained if we were to suppose that the cells gradually come to adhere so closely to one another that they tend to increase their surfaces of mutual contact. This would lead from the initial condition of roughly equi-dimensional cells loosely assembled together to a stage in which each cell is a thin flat disc attached over large areas to similar cells on each side of it. Finally, we might look on the eventual swelling of the cells into large bodies most of whose volume is occupied by fluid-filled spaces as a process which results in transforming a still greater proportion of the cytoplasm into external membrane in contact with the similar membranes of the neighboring cells (Mookerjee, Deuchar, and Waddington, 1953). We have recently examined the process with the electron microscope (Waddington and Perry, 1962; and Plate ii). In early stages of the neurula, quite large gaps occur between the cells of the presumptive notochord. These may possibly be to some extent exaggerated by the process of fixation, but there is no doubt that they indicate that adhesion between the cells is at that time rather low. As development proceeds, the areas of tight contact between adjacent cells increase, and the spaces between cells disappear. By the time the notochord cells have reached the "pile of coins" stage, the membranes between the flat disclike cells are in very close contact over their whole area. The membranes are slightly wavy or corrugated, but there is no important interdigitation of neighboring cells, nor can one find any well-formed desmosomes or attachment bodies. The evidence suggests that up to this stage there actually is the steady increase in the general adhesiveness of the cell surfaces which had been postulated earlier. At later stages, during the formation of the fluid-filled vacuoles, the areas of contact between neighboring cells suffer some remarkable changes. Large numbers of small more or less spherical vesicles appear in the neighborhood of the plasma membranes and become attached to them by tubular stalks. In some ways these vesicles resemble those which have been described at the margins of cells which are engaging in pinocytosis. However, as there is little or no fluid in the interspace between the closely apposed cells, pinocytosis in the ordinary sense of the term can almost certainly not be the explanation. Admittedly, the cells at this time must be absorbing fluid from somewhere in order to fill the large vesicles which occupy the central regions. In the notochord of the chick embryo at similar stages, vesicles opening on to the surface are found in the peripheral notochord cells which lie against the general body fluids, and it seems very likely that these are true pinocytosis vesicles at which the cells are engulfing fluid droplets (Jurand, unpublished). In the amphibia, however, similar vesicles are not found on the external surface of the notochord, which at this time begins to be covered with the chordal sheath. It seems rather more likely that the fluid to fill the central vacuoles gets into the cells by diffusion, and that the membrane vesicles just referred to are not directly connected with its uptake. They do, however, provide very definite evidence that the cell membranes are exceedingly active in some way or another at this time.

Another morphogenetic process in early vertebrate embryos which has been attributed to the activities of the cell membranes, is the rolling up of the neural plate into a tube (Brown, Hamburger, and Schmitt, 1941; Waddington, 1956). Here again, one of the first indications of the morphogenetic change of shape is an increase in the area of contact between certain cells; in this case, the cells lying near the outer rim of the neural plate. In these regions the neural epithelium, which was originally composed of more or less equi-dimensional cells, becomes transformed into a columnar epithelium, that is to say, the cells become long and narrow and, therefore, have greater area of surface in contact with one another.

The development of a columnar epithelium gradually spreads toward the center of the neural plate, and as it does so, the whole plate rolls up into a neural groove.

This process has not yet been fully studied with the electron microscope, but the data already available make it probable that something more is involved than simple increases in general cell adhesiveness (Selman, unpublished). Even at the blastula stage, the adhesion between ectodermal cells does not seem to be uniform over their whole surface. There is fairly close adhesion of neighboring cells in the region just below the external surface of the egg, but the more deeper lying parts of the cells are in only loose contact with one another. By the open neural plate stage, when the formation of a columnar epithelium is underway, the regions of adhesion between neighboring cells have begun to develop into structures definite enough to be referred to as terminal bars or desmosomes, although these are not yet nearly so well-formed as in later epithelia. By the time the neural groove is well folded up into a trough, one can find a fairly dense layer of granules lying just below the cortex in the region of the cells in which the attachment bodies are developed. This is very reminiscent of the terminal webs described by light microscopists in certain tissues (see Leblond, *et al.,* 1960).

The restriction of attachment bodies or desmosomes to particular regions of cells would give to those cells spatially definite properties for association with other cells. They might, thus, lead to the formation of organized groupings, rather as the directed valencies of atoms lead to their association into precisely arranged molecules. There are several instances of the orderly arrangement of small groups of cells which may find their explanation in this manner. One of the most remarkable of these relates to the two cells involved in the formation of the bristles in insects such as Drosophila. Each bristle really consists of a socket formed by a tormogen cell and a hair produced by a trichogen—there are one or two other pairs of cells which produce nerves and associated structures associated with these, but we shall neglect these for the present purposes. The hair which sticks out from the surface of the fly is secreted by the trichogen cell through a hole in the body of the tormogen. In a study of various mutants affecting bristles, a close correlation was found between the shape (thickness versus length) of the hairs and the orientation of the tormogen-trichogen pair (Lees and Waddington, 1942). Normally, the trichogen lies well below the tormogen, but in mutants in which the two cells lie more and more side-by-side, the hairs become thicker and shorter; while if the two cells lie entirely at the same level within the epithelium, they both give rise to sockets, and the hair in this case becomes completely transformed. In the past, it was very difficult to imagine how this precise spatial relationship between the two cells could be ensured, but it is perhaps not too difficult to imagine this if we can think of changes in the location of desmosomes or attachment bodies on the surface of the two cells in question. The situation has, however, not yet been studied with the electron microscope, so that one cannot say whether this suggestion is justified.

In other organs of Drosophila, we also meet very regular spatial arrangements of groups of cells. Once again, the ommatidium provides a good example. The basic structure is a group of four cone cells which lie above the eight retinula cells. These latter are arranged with great precision. The eighth retinula is a small cell at the base of the ommatidium. One of the seven cells has its nucleus at a slightly lower level than the other six which have their nuclei at the same distance between the external surface. The arrangement is actually somewhat more complex than this in the early stages, since from each cone cell a long process extends between two retinulae and reaches down to the bottom of the group of cells. As the ommatidium extends in length, these processes become thinner in cross section, eventually disappear, and can not be found in the adult eyes. Now, although within this group of cells no very well-developed desmosomes have been detected, there are indications of somewhat localized regions of stronger attachment between the retinula cells just where they abut onto the central space between the rhabdomeres (see Figure 14 and Plate xvii). It seems possible that further study, perhaps with other fixatives, will reveal more evidence of localized and specialized attachment areas. It certainly seems likely on general grounds that such a mechanism would provide the easiest way of building the cells together into such well-defined aggregates.

Another very general developmental phenomenon which might find its explanation in the development of specialized attachment bodies between cells, is the occurrence of polarized tendencies for morphogenetic movements within tissues. In the amphibian gastrula, for instance, fragments of tissue taken from the neighborhood of the blastopore not only have a tendency to elongate and to invaginate, but to do this in a direction corresponding to their original position in the egg. I was particularly struck with this when I did some experiments with the very rapidly developing egg of the anuran *Discoglossus pictus* (Waddington, 1941). In this the presumptive notochord area is originally widely ex-

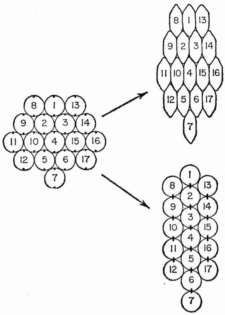

Figure 43. Possible mechanisms of anteroposterior elongation of gastrulating presumptive mesoderm in amphibia

On the left is the arrangement before gastrulation. If the surfaces of cell-contact running in the anterorposterior direction tended to enlarge at the expense of surfaces running from side to side, the cell group would be transformed into the grouping at the upper right. This would provide the required change in tissue shape, but at the expense of an alteration in cell shape, which probably does not occur. If, on the other hand, the cells before gastrulation have incipient attachment bodies or desmosomes at their anterior and posterior ends and if these gradually became joined up, the group would be transformed into the pattern shown at the lower right; and this would not involve any change in cell shape.

tended from side to side of the gastrula, and has to carry out, within a short time, a very great contraction in lateral dimensions and expansion in the anterior-posterior direction. Fragments of tissue taken from this zone exhibited a very powerful tendency to carry out this change in shape, and the polarity of the movement was firmly fixed within them.

Such tendencies could find an explanation if each cell in the tissue had a tendency to develop strongly acting attachment bodies on its anterior and posterior faces. This would cause the cells to become lined up in the anterior-posterior direction, thus causing an elongation of the whole tissue mass along this axis (Figure 43). Once again, however, this is a speculative suggestion which has not yet been verified. It is mentioned here to draw attention to the fact that the discovery of attachment bodies and desmosomes provides us with some quite new lines of approach to the old problems of tissue polarity and cellular arrangement, which have offered a challenge to embryological thought for such a long period.

In general, the newer ideas about cell adhesions, whether these are thought of in terms of general physical attractions, such as those discussed by Curtis, or as involving more localized attachment bodies, have given us a considerable repertoire of new ideas with which to attempt the explanation of many processes of early morphogenesis, in which relatively small numbers of cells are involved. When we are dealing with tissue masses made up of many thousands of small cells, forces inherent in the plasma membranes seem perhaps less suitable for accounting for such over-all properties as we may discover. This, however, is the problem to which we will turn in the next section.

In Tissues

By far the greatest part of experimental work on morphogenesis has been conducted on masses of tissue which contain many more cells than those which have just been discussed. It is in such complex multicellular masses that the problems of the arising of new patterns during development are exhibited in their full richness and difficulty. Unfortunately biologists have not met the challenge by developing in the last few years an equivalent number of new patterns of thought. In spite of the volume of older work on the morphogenesis of tissue masses, I shall treat it rather briefly in this chapter.

It will be as well first to look at an example which brings before us clearly the type of problem with which we are faced. In general terms, this problem is that masses of material, which are biologically speaking highly complex, consisting of many cells often of somewhat different types, nevertheless become molded into structures which have shapes which are not only definite and repeatable from one individual to the next, but which, in some cases at least, exhibit considerable geometrical simplicity. A clear and simple example is provided by an old investigation of Anikin (1929) on the formation of the cartilages of the digits of the embryonic amphibian limb-bud. These cartilages form as condensations within the mass of mesenchyme with which the early limb-bud is filled.

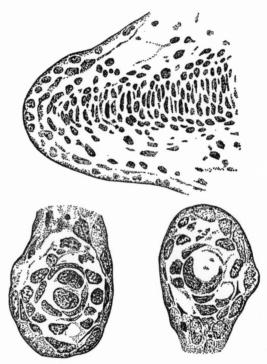

Figure 44. Shapes of nuclei in precartilage cells

A longitudinal section (above) and two transverse sections (below) through a toe of a developing foot of the newt. (After Anikin, 1929)

In the distal region of the limb, five condensations appear, corresponding to the five fingers or toes. Each condensation is quite small; in fact, in a short time the pre-cartilage cells for each digit arrange themselves in a single row. Within each of these small cylindrical units the constituent cells exhibit a further degree of geometrical regularity. This is expressed particularly in the shape assumed by the nucleus. In cross sections of the cartilage, the nuclei show a variety of kidney or lunate shapes; the more kidney-shaped ones lie nearer the mid-line, while the peripheral ones are more elongated (Figure 44). Anikin showed that the whole series of these shapes could be described by a relatively simple geometrical formula relating the shape of the nucleus to its distance from the central axis. In his formulation, a nucleus initially lying just to one side of the central axis is imagined as being pushed out in a radial direction. The distance through which each part of the nuclear periphery moves is supposed to be inversely proportional to its distance from the central axis, so that as the nucleus recedes from the center, it becomes more lunate in shape. This gives a reasonably accurate geometrical description of what we see (Figure 45). The difficulty is to know how to interpret it. It is not at all clear that any such translation of the nuclei actually occurs; even if it did, it would not be easy to imagine what forces could produce the motion.

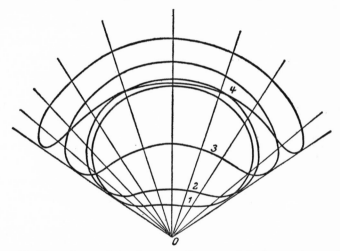

Figure 45. Transformation of precartilage nuclei shapes in movement from central axis (After Anikin, 1929)

The whole scheme is, indeed, best thought of as no more than a way of formulating a set of geometrical relations.

The case studied by Anikin is a particularly simple and clear-cut example of a general class of phenomena, in which some end result which appears relatively simple is attained but it is not easy to discover any equally simple causative agents. In general, biologists have not been able to progress any further in understanding such matters than to postulate that there exists a morphogenetic field which controls the shape of the structure produced. The word "field" is, of course, a very vague one. In its most definite usage, it would imply that within the region of a developing structure some force (or simple set of forces) is operating and exhibits a regular spatial distribution in intensity. The difficulty in the biological cases is to determine what this force might be. When the resulting structure is geometrically simple, so that it could be described by a fairly elementary algebraic function as in the case described by Anikin, the attempt to identify a field force might appear reasonably hopeful; but in many cases, of course, the shapes involved in biological structures, although quite definite and precise, are not geometrically at all simple. Consider, for instance, the bones of the limb: each has a definite shape and is precisely formed even in detail; and each is provided with regular joint surfaces where the two bones come in contact, and so on. It might seem likely that these joint surfaces are formed under the influence of the function of the limb as the bones are moved by the attached muscles, and that we do not need to find any other causative agents to bring about their morphogenesis; but this is not the case. Well-developed limb skeletons with adequately formed joints are produced in nonfunctioning limbs (Hamburger and Waugh, 1940). The factors determining their morphogenesis must arise within the mass of developing cells, independently of function. It is very difficult to imagine what such field forces might be. The difficulty is increased if we remember something of the complexity of the processes by which the shape of a bone is in fact molded; it is produced by an elaborate process of deposition of bone material in certain parts, accompanied by its readsorption in other parts. A few years ago in our laboratory, Bateman (1954) took advantage of certain mutant genes in the mouse to investigate these processes of adsorption and deposition in detail. In microphthalmic and gray-lethal mice there is a slowing down of bone accretion and a lack of bone erosion. By comparing a series of stages of bones from such mice with those from normal animals, it was possible to show the great local specificity of these two processes within a normal developing bone rudiment (Figure 46). In a system which is actually behaving in this way it seems rather hopeless to try to search for a single unitary field force to which the whole morphogenesis can be attributed. The final end result we see as the shape of the adult bone or other organ must actually, in spite of any apparent simplicity it has, be produced by the cooperative effort of large numbers of different processes.

Probably the best case to illustrate this point in detail is still the Drosophila wing, although most of our understanding of the morphogenesis of this system dates from some twenty years ago (Waddington,

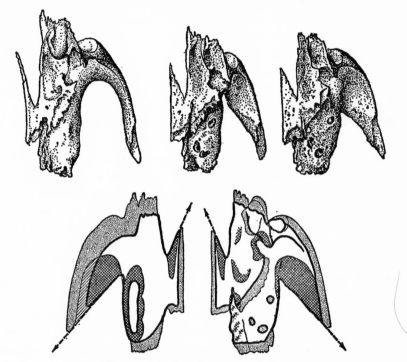

Figure 46. Development of bones

At top are drawings of the maxilla of the normal mouse, of gray lethal, and microphthalmic mice. At bottom are diagrams (inferior aspect on left, superior aspect on right) showing areas of deposition by dots, areas in which bone is first deposited and later eroded by hatching, and areas of erosion by cross hatching. (After Bateman, 1954)

1940b; Henke, 1947; Stumpf, 1959). The wing is a rather simple organ, consisting essentially of an epithelial sac which has become flattened into a bilaminate blade of a more or less oval and quite definite general shape. From the root of the wing blade to the tip, five radial veins diverge and are connected by two cross veins. This final shape is the resultant of a number of component processes, which can often be identified by observing the development of mutant types. One of the first steps in the development of the wing from the imaginal bud is, for instance, the folding out of part of the epithelium of the bud into a protruding sac. The position along which this fold takes place determines the shape of the margin of the wing, and is affected by a large number of genes of which *vestigial* and its alleles are some of the best known. The initial protruding fold then expands in area by the thinning out of the epithelia of which it is composed, the cells becoming converted from a columnal to a flattened shape. During this stage (the pre-pupal instar), cell division also proceeds. This process does not take place in a random fashion. The mitotic spindles are orientated, and this brings about differential growth of the rudiment in various directions, which can also be affected by mutant alleles, such as those of *broad* and *narrow*. At the end of this instar, the wing rudiment becomes inflated by the pressure of hemolymph within it, so that it is converted into a thin-walled inflated sac. At the beginning of the true pupal instar, this sac contracts, the hemolymph being forced out of it again, and the upper and lower surfaces of the wing coming into contact. The degree of contraction which is achieved, is one of the most important factors determining the final shape of the adult wing; and it is the resultant of a balance between forces due to the contracting epithelium, the pressure of hemolymph, the existence of the thickened rudiments of the final veins, and the shape of the wing margin, all of which can be altered by the presence of abnormal mutant alleles. In stages just subsequent to this a series of other processes determine the precise pattern of development of the veins themselves. Then follows a stage in which the individual cells making up the wing epithelia expand in area, a process affected by genes, such as *miniature*. Finally, after the animal emerges from puparium, the wing blade, which has been crumpled up during the last hours of pupal life, becomes unfolded and dries out. Only if the drying occurs equally on the upper and lower surfaces does the wing lie flat; otherwise, it becomes bent as in *Curly* or *curved*.

In this summary, only the morphogenetic processes which have powerful effects on the general shape of the wing have been mentioned. Even so, one can see that the relatively simple adult shape is the result of the cooperative action of quite a large number of interacting processes. If one considers the wing as a whole, including the pattern of veins within it, about a score of such processes have been identified. In such a situation it is clearly inappropriate to look for a single overriding field force. The morphogenetic field is a region in which many forces are at work. One can think of the development of a structure of this kind as formally analogous to Rugby football. The ball may be dribbled in a particular direction by the combined action of six or seven forwards; a small kick first from one, then from another, then from a third until it is impossible to identify any one particular individual as determining its course.

Situations of this kind present us with two different types of problem. The first is to try to discover the particular causative agents which are at work. The result of this will be to identify not *the* field force, but only one or a few out of the many forces which make up the whole field system. These forces will in general be produced by the global properties of masses of cells, rather than arising directly from the intracellular processes which are under immediate gene control.

In the Drosophila wing, one of the active factors is the contractility of the epithelium in the contracting wing, and another factor is the pressure of fluid within it. Tuft (1962) has suggested that fluid pressure within spaces in developing organ systems is often a very important morphogenetic force. In the early embryonic development of amphibia, for instance, we see first the formation of a blastocoel filled with a watery solution, and after a time this disappears and a fluid-filled archenteron is formed. Water actually accounts for almost three-quarters of the volume of the young amphibian embryo, and many of the changes of shape which occur during early development can be explained in terms of the uptake and redistribution of water. From studies on the density of the embryo, of its weight in water, and on the osmotic pressure of the fluids in its various cavities, Tuft was led to the conclusion that the various regions differ in their capacity to bring about a net flow of water through them. In the earliest stage, for instance, there is a net flow of water inward through the animal pole material of an amphibian egg, and a net flow outward through the vegetative material. The inflow is at first the larger of the two, and this leads to the growth of the blastocoel cavity. With the gradual extension of the area derived from the vegetative cells during the process of invagination, the outflow gradually becomes larger, so that the fluid is eventually removed from the blastocoel. When the blastopore closes, the water passing outward through the endodermal surface becomes trapped in the archenteron, which becomes swollen into a hollow cavity. Many other morphogenetic events, such as the swelling of the neural tube, the formation of the eye cups, may also be attributable to the movements of water from one part of the embryo to another. It is to factors of this kind, which are very complex secondary resultants of the basic gene-action systems, that we have to look to identify the operative forces in the morphogenesis of multicellular systems.

The second set of problems which arise in this connection are those concerned with the over-all direction in which the development of the field takes place. A Rugger pack, to return to that analogy, dribbles the ball in a definite direction, usually toward the opponents' goalline. The global concept of the general direction in which the field develops becomes interesting when we can find means of controlling it. In the biological case of a developing organ system, this can sometimes be done. As D'Arcy Thompson pointed out (1942), something of this kind has certainly occurred in evolution. His well-known demonstration that the shapes of certain species, for instance, fish, can be transformed into the shapes of other species by a general deformation, is an example of the diversion of a general global direction of development of a field. Many other similar examples have been described since he wrote. In this context, however, we are still very much at the level of merely describing empirically produced phenomena. It is one of the areas in which much more abstract theoretical thought is called for to clarify how best to formulate the over-all regularities which can be detected in the behavior of field systems. Medawar (1950) has made some efforts in this direction but much more is required. One has the feeling that some new pattern of thought should emerge, but it has not yet done so.

In order to emphasize the potentially very great importance of an understanding of these over-all results of field systems for general biological theory, I shall mention shortly two or three recently described examples. These relate to the effects of adding to or subtracting from the quantity of tissue within which the field system is operating.

When the mass of tissue developing into a given organ or organ system is reduced below the normal size it may turn out that one is dealing not with a single unitary system but with one which is composed of two or more subsystems. For instance, Lehmann (1948) has shown that when the amount of neural plate material developing into the brain of an amphibian embryo is reduced by successive amounts, the first effect is to reduce only the forebrain, leaving the mid- and hindbrain unaffected. It is only after the subfield of the forebrain has been reduced to a small or negligible amount, that the more posterior parts of the brain become affected.

Hampé (1959) has described a similar situation in the chick wing. In this system, there is normally a competition between the rudiments of the tibia and fibula for the available supply of mesenchyme cells. If the size of the rudiment is reduced, all the cells may be taken over by the tibia, which appears in full size, while the fibula does not develop at all. On the other hand, competition can be prevented within a rudiment of normal size by inserting a plate of mica between the two early rudiments, and in this case the fibula develops into a larger size than usual since its cells cannot be attracted away from it by the rapidly developing tibia. Similarly, if extra cells are grafted into the early rudiment the fibula becomes larger than usual.

When a given field system is effectively increased or decreased, the effect is usually not simply to produce an organ of enlarged or diminished size with normal morphology; instead there may be an actual change in the details of the structures which are formed. For instance, in the chick wing bones mentioned above, Hampé found that, as the size of the fibula rudiment increases, its distal end becomes more fully formed until eventually it forms a well-developed fibulare (Figure 47). This is a bony structure which is absent in modern birds but found in their earliest fossil ancestors, such as Archaeopteryx. Thus, some of the changes in morphology which have taken place in evolution may have been brought about by simple alterations in the quantity of material available for the operation of certain field systems (Figure 48). Other discussions of these very interesting phenomena will be found in Wolff (1958) and Kroeger (1960).

Although in the chick wing both the rudiments of the tibia and the fibula appear as simple condensations of mesenchyme, the phenomena just described show that they behave as specifically different entities. It is remarkable how soon the different structures appearing within a given field system seem to acquire their own individuality. For instance, the different toes within a developing amphibian limb respond differentially

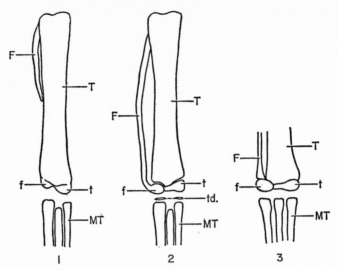

Figure 47. Results of adding material to the developing chick limb-bud

In 1 is the normal skeleton of the chick leg; 2, that developed by a leg-bud to which extra material had been added at an early stage; 3, that of the fossil ancestral bird Archaeopteryx. Note the enlargement of the fibula (F) in 2 and its connection with the fibulare (f), to give a condition similar to that of Archaeopteryx. (After Hampé, 1959)

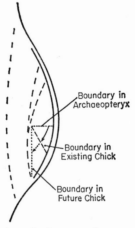

Figure 48. Illustration of evolutionary changes in morphology of avian leg skeleton caused by changes in proportion of presumptive material in various bones

The lines indicate positions of the boundary between the tibia area and the fibula area. (After Hampé, 1959)

to general inhibiting agents. Tschumi (1953) inhibited the development of Xenopus limb-buds with chloroethylamine and found a definite pattern of sensitivity, the toes being affected in the order 1, 2, 5, with 3 and 4 persisting to the end. The fact that during evolution particular toes of the pentadactyl pattern have been increased or decreased in size relative to the others, is again evidence that each toe has its own individuality, and can be affected more or less specifically by genetic factors. It is not the case that the pentadactyl pattern can only be affected as a whole according to some general scheme of modification (Figure 49).

We are still, I think, quite ignorant as to the nature of the individual specificities of such organs. They must ultimately be chemical in nature, but the kind of the chemical differences involved is still extremely obscure. The problem with which we are faced is exhibited very strikingly in some results from experiments by Saunders, Cairns, and Gaseling (1957). Mesoderm from the thigh region of an early hindlimb rudiment in the chick was grafted into the distal region of the forelimb. In this new location it developed into a distal part of the limb, namely, into digits; but in spite of the fact that it now lay within the forelimb and had adopted the proximo-distal character appropriate to its position within that limb, it nevertheless retained the character of being hindlimb material. It developed, in fact, into a claw attached to the distal end of the wing. This case seems to demonstrate that there is some characteristic, presumably chemical in nature, which belongs to the hindlimb as a whole and differentiates it from the forelimb as a whole. It seems impossible at present to guess what kind of chemical difference might be involved. It should, however, be possible to investigate the matter either with immunological methods or perhaps by following the specific passage of macromolecules incorporating radioactive tracers from grafted organs into the corresponding organs of the host.

In general, we have as yet hardly any ideas about the chemical nature of the substances involved in the activities of field systems. There is, of course, a lot of evidence showing that one part of the tissue in a developing organ may influence another part. Most of this evidence, however, remains at a strictly biological level, and only very rarely can we say anything about the nature of the material causative agents involved. The biological evidence is very interesting in itself.

A rather special example of mutual interaction between fields was

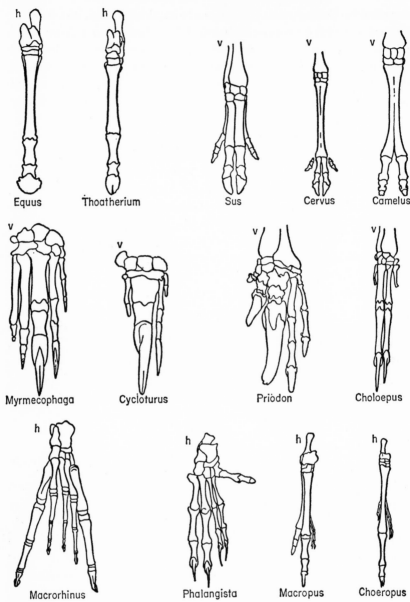

Figure 49. Modifications of basic pentadactyl pattern in evolution (From Tschumi, 1953)

described by Goldschmidt, Hannah, and Piternick (1951). In a strain of Drosophila which had been selected to show various abnormalities in the eversion of the imaginal buds of the limbs, some cases were found in which two forelegs had become completely unified. The combined mass, which should have contained two complete leg field systems, developed into a unified organ of rather regular shape, illustrated in Figure 50. As

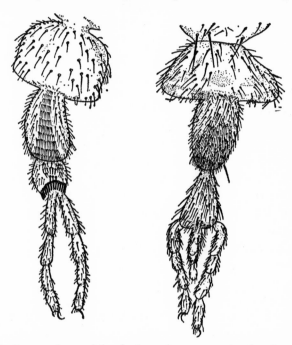

Figure 50. Organs formed by fusion of two anterior leg imaginal-buds in Drosophila (From Goldschmidt, Hannah, and Piternick, 1951)

Goldschmidt points out, one could quite well imagine an organ of this kind having, under some circumstances or other, turned out to be of natural selective advantage, so that it became stabilized as a characteristic of some taxonomically defined group of animals. The union of two separate fields and their common development into a synthetic combination of the two originally distinct parts, is one of the processes which may have been involved in the divergence of the major groups of animals. At least it provides an interesting illustration of the types of flexibility which exist within a developing system and which must limit or define the paths along which evolutionary change can easily occur.

The usual types of tissue interaction during the development of a field system involve processes of induction, in which one set of tissues acts as an inducer which causes the formation of corresponding organs in neighboring tissue competent to react to it. One of the classic and most fully studied examples of this is provided by the work of Baltzer and his pupils (Baltzer, 1952, 1957; Henzen, 1957; Wagner, 1949) on the formation of the organs of the head in amphibia. In these experiments pieces of tissue taken from an anuran larva were grafted into a urodele and *vice versa,* so as to give tadpoles which had combinations of tissues from the two sources. The main anatomical elements involved in the system are the internal tissues, which form cartilages derived from the mesoderm, and the external ectodermal organs which cover these. Both sets of organs are quite different in the two classes of amphibia. The head skeletons of the toad (anuran) and the newt (urodele) used by Baltzer are illustrated in Figure 51. It will be seen that there are considerable differences in the shape and arrangement of the cartilages; in particular the toad contains infrarostrals which do not occur in the newt. Again, on the external surface of the head is the difference that toad tadpoles have suckers while newt tadpoles carry thin cylindrical protrusions which are known as balancers.

When grafted foreign mesoderm gets into the head region, it forms cartilages and bones which are, at any rate roughly, in the right positions in relation to the main part of the host's head (Figure 51). They cannot, however, be in exactly the right positions, because toad cells, even in the head of a newt, form characteristically toad cartilages, while in the reciprocal combination newt cells form newt cartilages in the head of the toad tadpole. Usually the two taxonomically different types of tissue separate from one another. Occasionally one may find a bone which contains both newt and toad material, but this happens only in rather featureless parts of the skeleton; newt material never forms the characteristic toad rostral cartilages, nor do toad cells ever contribute to the formation of the bony dentale which is characteristic of the newt. However, the fact that toad cells in a newt head will develop into a rostral in approximately the correct region, even though the newt itself has no cartilage corresponding to this in its head skeleton, shows that there must be some very general similarity between the head fields in these two classes of animals. Even in the newt there must be some sort of signal in the region which corresponds to the toad's rostrals, and this signal can be, as it were, recognized by toad cells, which respond by forming a cartilage in that place. We can express this by saying that the induction

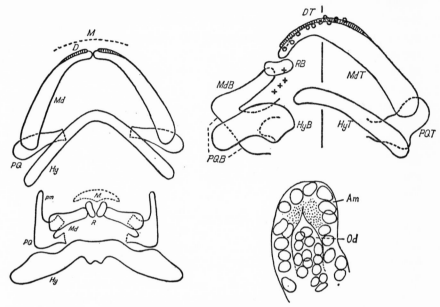

Figure 51. Amphibian head skeletons

On the left are ventral views of the head skeleton in a newt larva (above) and in a toad Bombinator (below). (M), future position of the mouth; (Hy), hyoid; (PQ), palatoquadrate; (Md), mandibular; (R), infrarostrals, only present in anura; (D), dentale, only present in urodeles. Above on the right is the head skeleton of a newt larva into which toad cells had been implanted on the side which is left in the figure. Note the formation of rostrals (RB) by the toad cells, and in general their production of toad-like structures. Circles are newt teeth, crosses chimerical teeth. A section of one such tooth is shown below on the right. The ameloblasts (Am) are from the newt host, the odontoblasts (Od) from toad material. (After Baltzer, 1952, 1957)

field is very similar in the two classes, but that the competence to react is highly species-specific.

Much the same situation occurs with respect to the external organs. Many years ago Spemann and Schotte (1932) showed that ectoderm from one species grafted on to the underside of the head of the other species, forms the external organs appropriate to its own genetic character, that is to say, toad ectoderm forms a sucker, while newt ectoderm forms a balancer. These organs are located at slightly different parts of the head, but in the mosaic produced by grafting, they appear in their own appropriate places (Figure 52). Thus, in the region where the toad tadpole bears a sucker, the interior organs of the newt head cause toad ectoderm, if it is present, to produce that organ. Similarly, in the part of the toad corresponding to the balancer position in the newt, there must be something to which newt ectoderm can respond by developing into a balancer.

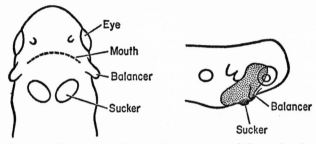

Figure 52. External structures on amphibian heads

On the left is a composite diagram showing the approximate positions of the newt balancers and the toad suckers on the ventral side of the tadpole head. On the right is a drawing of a newt tadpole developed from a gastrula on to which toad ectoderm has been grafted. The toad material (shaded) has formed a sucker in the region where toad tadpoles normally bear suckers, and is also taking part in the formation of a balancer in the place proper for balancers on newt tadpoles. (After Baltzer, 1952)

The internal organs which induce these foreign suckers or balancers are presumably not the same as the structures which induce them in undisturbed tadpoles. This is perhaps best demonstrated in connection with the formation of the mouth. It has been shown by other experiments that in the toad tadpole the formation of the mouth is normally induced by the rostral cartilages. These are absent in the newt head; yet, if toad ectoderm comes to lie in the mouth region of a newt embryo, a mouth will be induced to form in it. The induction seems, probably, to be done by the central part of the mandibular cartilage, since if this is displaced the mouth may be displaced. The mouth that forms is a typical toad mouth, that is to say, it is considerably smaller than the mouth that would be formed by the newt embryo itself out of its own tissues.

The results in all these cases can be expressed by saying that the head is pervaded by a field system which is extremely similar in the two classes of animals, and that various organs are produced by the responses which the available competent tissues can make to these fields. These reactions are in general limited by the specific nature of the competent tissues. It is a question whether cells from one species, such as the toad, can ever be persuaded to take part in the formation of some organ which they do not normally produce. There are few indications that this ever happens except in connection with the formation of relatively undifferentiated and simple structures. For instance, when newt ectoderm placed on the surface of a toad head is developing into a balancer, some toad mesoderm cells may be found within the cylindrical out-growth organ. The internal tissues of the balancer are, however, a rather simple type of mesenchyme, and it does not seem likely that any very specific synthetic processes are necessary to enable toad cells to take part in such a comparatively un-differentiated performance. Similarly, if chick ectoderm is grafted over the early leg-bud of a duck embryo, it will take part in the formation of webs between the toes, which of course are absent from the chick's own legs (Hampé, 1959); again the differentiation involved is rather slight. The nearest one gets to finding cells in these graft mosaics doing something outside their normal repertoire, is the fact that toad mesoderm cells can take part in the formation of newt teeth, although the toad tadpole itself does not produce any teeth (Figure 51). However, after metamorphosis teeth do form in the mouth of the adult toad; the participation of toad cells in the development of teeth in larval newts is perhaps more in the way of an anticipation of what they can do in later life, rather than the performance of something quite foreign to their developmental repertoire.

Toads and newts are, of course, biologically only rather distantly related to one another. We cannot perform any genetic analysis of their differences and cannot prove that it is the genes, rather than, for instance, the cytoplasm that limits the competence which the cells can show when grafted. On general grounds, however, this seems most probable. Moreover, an essentially similar situation can be discovered in organisms where a genetic analysis is possible. For instance, Stern (1954, 1956; Stern and Hannah-Alva, 1957) has studied the formation of bristles and sex-combs in Drosophila. Although it is not technically possible in this organ to make grafts comparable to those between newt and toad tadpoles, genetic stocks are known which encourage the formation of mosaics in which patches of genetically different tissue are formed within the body of one and the same individual. The mosaic patch of genetically altered tissue gives much the same kind of information as the grafts just described.

Investigations revealed a situation exceedingly similar to that which we have discovered in the head cartilages of the amphibian tadpoles. The appearance of a particular organ depends on the presence of a general field system which is responded to by the available cells according to their limited competence, and in this case, it could be shown that the competence is determined by their genetic constitution. For instance, in the foreleg there is a field system—which Stern refers to as a "pre-pattern"—which can cause the production of a sex-comb in the appropriate region if any tissue competent to develop into this organ is present. This field is the same in females, which do not bear sex-combs, as in males which do. If in a female leg a small piece of male tissue is formed as a mosaic patch in the sex-comb region, then it will develop into a small piece of sex-comb. This is extremely similar to the behavior of a group of toad cells which form a rostral cartilage in the appropriate region of a newt head. Again, on the thorax of the fly there is a field system or pre-pattern concerned with the formation of bristles, and the actual bristles which develop depend on the competence of the tissues which are present to react to this system.

As the situation in the tadpole head shows, during the course of development there is a sequence of stages in which pre-patterns become developed into definitive patterns which then in turn act as pre-patterns for the next stage. If we begin at the stage where there is an inducing field or pre-pattern within the mesoderm of the head, this reacts with the competence of the available tissues to produce a certain definite pattern of cartilages or bones; but these then proceed to act as an inducing field which controls the formation of the next pattern to appear, that of the ectodermal head-organs, such as the mouth, suckers, or balancers, which then constitute a second-order definitive pattern. In this case, the first- and second-order patterns may well extend over the same total area. Often, however, the sequence of developmental stages involves the breaking down of an extended first-order pattern into a set of more restricted separate second-order patterns. For instance, it is most probable that in the above example, the mouth soon comes to act as an inducing field for the minor details which will develop within it; and in doing so becomes independent of, for instance, the suckers, each of which will also constitute a separate field controlling its own details.

The succession of fields and patterns and their fragmentation into sets of separate subregions, is not well exhibited in Stern's studies on bristle patterns in Drosophila mosaics, but the same principles certainly apply to that organism. Their operation can be seen, for instance, in the development of flies homozygous for various mutant alleles of *aristopedia*. This gene causes a leglike organ to be formed in the place normally occupied by the arista. It is not clear whether this change should be considered as resulting from an alteration in the field controlling the character of the imaginal buds, which converts the antennal disc into a leg disc; or to an alteration in the competence of the tissue, so that the material in the antennal bud reacts abnormally to its situation and develops into a leg. The fact that in some strains we find organs which are part leg and part arista and that the proportion between the two can be altered by temperature treatments in quite late larval life, could be interpreted in either way, though perhaps rather more easily in the latter (see Waddington, 1940c; Vogt, 1947). But, whichever way we may like to interpret this primary pattern change of arista into leg, a second step succeeds it; and this is also brought about by the *aristopedia* alleles. Many of these have an effect on the segmentation of the tarsus of the leg; and when the antennal leg is well enough developed for the effect to be verifiable, it is found also to exhibit the same aberrations as the main legs (Waddington and Clayton, 1952; and Figure 61). It is scarcely possible to interpret this except as an altered reaction of the tissue to normal pre-patterns within the legs, whether main legs or antennal legs.

Similarly the genes which determine the number of tarsal segments which will be formed in a Drosophila leg, which we shall discuss in the next chapter, can be regarded as altering the competence of the tissues to react to an inducing field or pre-pattern which specifies that a leg is to be formed; and when this primary pre-pattern has itself been changed so that a leg is produced from the antennal disc, the altered competence operates in that location also, giving shortened antennal legs. Once again the final adult form is a consequence of a series of determinative steps of pattern formation, in which the pattern of an earlier stage (e.g., the determination that here there shall be a leg, there a wing, here an antenna, and so on) functions as a pre-pattern or inducing field at a later stage, to which the tissue in the leg region, for instance, responds by developing, according to its own constitution, either a five-segmented or a four-segmented pattern; and then this pattern in its turn acts as a pre-pattern, to which, for instance, male or female tissue will respond by developing sex-combs in the appropriate locations.

It is clear that we do not get very much further in seeking an explanation for the general phenomenon of pattern formation, merely by saying that it is all due to pre-patterns to which competent tissues react. If we discover that the production of a balancer at a particular place in a tadpole, or a hair at a given location on a Drosophila thorax, is due to an interaction between competent tissue and inducing field, this is interesting information about the developmental mechanics of the balancer and the hair, but not about the way in which structural order comes into being. The real problem we have to tackle is, How does a pre-pattern or inducing field arise? In more general terms, the essential problem is that of increase of complexity of pattern. Suppose we begin with a pre-pattern (or inducing field, or template) which contains the specification of n definite positions; and that we find at a somewhat later stage of development that $n + m$ entities have been formed in precisely defined places; How has this been brought about? Clearly the problem reduces to that of a complex response to a simple stimulus. At some of the n places specified by the original pre-pattern, more than one entity is developed; and these entities are grouped in some orderly arrangement, so as to produce the $n + m$ defined positions we have postulated. The conundrum which confronts us is not merely one-many, but one-ordered-many. To the single specification "head," the mesoderm of the tadpole responds by the formation of a number of different cartilages in definite places; to the determination "leg," the competent Drosophila tissue responds by forming an organ with either five, four, or some other number of segments with the property of producing sex-combs at a definite place if suitable material is available; and so on. In all these cases it is possible that there are actually several stages, which we may not yet have distinguished, between the single specification (such as head) and the relatively complex arrangement which we take to be the response. But the mere invocation of some intermediate stages does not do away with the fact that somewhere along the line an increase in complexity occurs.

Conclusion

This problem of the increase in complexity is the most mysterious of all those with which we are faced in our investigations of morphogenesis. We have seen that where small groups of cells are involved we already have certain lines of thought (most of them rather new) which hold out the hope of explaining some of the phenomena. Interactions of cell membranes and the formation of attachment bodies or desmosomes, give rise to mechanisms which may operate in a wide variety of situations. But when we come to consider morphogenesis in masses containing large numbers of cells, our means of understanding are much less powerful, and it is often difficult to resist the temptation to talk in terms of vaguely defined concepts, such as fields. But a field has really no explicative power unless we can say what it is a field of. As yet we know extremely little about the material nature of the causative agents which are active in the situations where we are tempted to speak of fields. Is the position of the head cartilages within the mesoderm of an amphibian tadpole, or the position of the sex-comb within the Drosophila leg, determined perhaps by the diffusion of various substances from centers within the mass? If so, are these substances of a macromolecular kind? Or are we perhaps concerned with the distribution of grades of intensity of metabolic processes involving small molecules? Or are we confronted with the results of some type of physical variation?

We have as yet only very slender clues. One is the argument that, since the inducing field can cause the development of organs which are quite foreign to it, such as rostral cartilages within a newt head or a balancer on a toad tadpole, it is unlikely that any very specific substances are essentially involved. To put it in another way, the information coming from the underlying tissues in the toad head which causes a newt balancer to appear, can scarcely be very great, since the character of the balancer is determined, not by the transmitted information, but by the reaction to it. Another hint can perhaps be found in the recent studies on the capacity of various tissue extracts to induce different regions of the central axis of the embryo in amphibian gastrula ectoderm. It has been shown, particularly by Yamada (1961), that a certain nucleo-protein fraction, which when freshly prepared induced mesodermal structures and the posterior region of the trunk, can be converted by successive short heat treatments into material inducing progressively more anterior regions of the body. Yamada has suggested that in general it is the protein, rather than the nucleic acid part of these fractions, that is active. It seems rather probable, therefore, that comparatively slight changes in the physical configuration of protein molecules may have a profound importance on the regionality of the material which they in-duce. How these results can be applied to the general consideration of the active agents in inducing fields, must be for the future to determine.

Apart from such vague and not easily interpreted indications, we re-main almost entirely in the dark. There would appear indeed to be much comparatively simple experimentation waiting to be done in this field. For instance, we do not know what would be the effect of separating parts of the field by micropore filters of various permeabilities. Again, no tracer experiments appear to have been done in this connection. The resources of immunology have also not been called on. One gets the impression that in this connection, as in that of the specific nature of complex organs referred to earlier, there may be a whole gamut of biologically active substances waiting to be discovered.

6. Biological Patterns

I *AM* using the word "pattern" to refer to complex forms in which I wish to distinguish a number of different parts and to discuss the relation of these parts to each other. There is, of course, no absolute distinction between the structures which I choose to refer to as patterns, and those which have been considered in the earlier parts of this book, since any form has a pattern aspect which can be taken as the center of attention if wished. There are, however, a number of biological phenomena, such as the colored patches on a butterfly's wing or the bristles in their orderly arrangement on a fly's thorax, which it is conventional and convenient to refer to as patterns. Since some of these cases are good examples of certain general principles in morphogenesis, they are worth discussion at some length. I shall arrange the material under three subheadings: All-Over Patterns or Textures, Area or Volume Patterns, and Spot patterns.

All-Over Patterns or Textures

The configurations of the surfaces of biological systems become interesting from the point of view of morphogenesis only when they exhibit at least a moderate degree of orderly arrangement. An example of this kind, which has been fairly fully studied and which exhibits some properties of general interest, is the texture of the adult cuticle on the abdomen of the bug *Rhodnius prolixus*. In larval stages, each segment of the abdomen bears a large number of hairs. When the adult instar is reached, the hairs are replaced by a uniform-looking series of transverse ridges. Locke (1959) has made an extensive series of experiments in which small fragments of larval skin were cut out and grafted either into the same or another location and in various orientations, and he observed the result of these experiments on the ridge pattern in the adults.

The experiments showed, rather surprisingly, that the ridges within a segment are not all equivalent to one another, but have a considerable degree of individuality. Grafts in which the levels along the antero-posterior axis are unchanged give rise to normal adult patterns; but if the levels along this axis are altered, the pattern of the adult is affected. The adult pattern, indeed, develops quite normally if a fragment from one

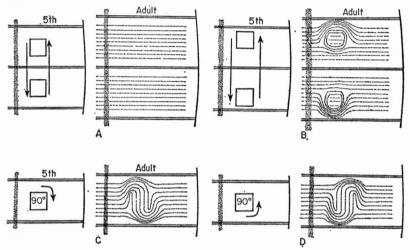

Figure 53. Effects of grafting larval (5th instar) skin in Rhodnius on cuticular pattern on adult abdomen

A, B, grafts from one segment to another give normal patterns provided the level within the segment is not altered. C, D, grafts rotated through 90° (or other angles) give patterns depending on the direction of rotation. (After Locke, 1959)

segment is grafted into a segment in front or behind its place of origin as long as it is placed in the same position within the segment, and the same occurs if it is transplanted from the right side of the mid-line to the left or vice versa. If, however, the position of the fragment within the segment is altered, then the adult ridge lines run round the graft as illustrated in Figure 53. Disturbances of the pattern are also caused if a fragment of larval skin is cut out and replaced after rotating through 90°.

All these results can be explained if we suppose that each ridge has a certain value on an anteroposterior gradient of some kind. If an area of skin is moved forward or backward within a segment or is rotated, the result will be that the potential ridges within the graft no longer agree in value with the potential ridges in the surroundings (Figure 54). The rule is that when the host ridges are more anterior in character than the graft ridges, they run forward around the front end of the graft, while when the host is more posterior its ridges run posteriorly around the grafted area.

The significance of this example is that it provides another instance of the acquirement, by elements within a pattern which look quite similar to one another, of an actual individuality which one would at first sight not have suspected. This is a reminder of the way in which the different

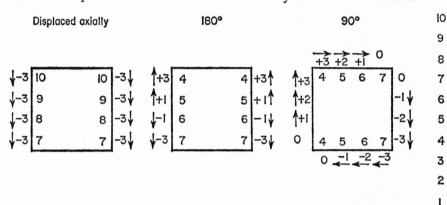

Figure 54. Hypothesis of skin grafting effects in Rhodnius

Each segment is supposed to be characterized by an anteroposterior gradient indicated by the figures from 10 to 1 on either side. On the left is a square which has been displaced posteriorly; the other two drawings show squares which have been rotated. The figures outside the squares show the difference in value between the host tissue and the neighboring part of the graft; the host ridges will run forward if this difference is positive, backward if it is negative. (From Locke, 1959)

cartilages within a developing limb very early take on some specific distinct characters of their own. Once again it appears that the chemical basis underlying the phenomena of morphogenesis are much more complex than might appear.

The only other all-over pattern which I wish to discuss is one which provides a beautiful example of the general principle that development often involves alternative end states between which the developing system can easily be switched. The system I have in mind is the color pattern on the fresh water snail *Theodoxus fluviatilis* studied by Neumann

(1959a, b). In this snail there are three basic color patterns. In one, the sector pattern, there are alternate stripes of pale and dark color running longitudinally along the length of the helically coiled shell. The second pattern is cross-banding, in which the dark and light stripes are formed parallel to the lip of the shell. A third pattern, known as the drop pattern, is formed by longitudinal stripes of constant width, each stripe consisting of alternate dark and light patches (Figure 55). These patterns

Figure 55. Patterns on shells of Theodoxus fluviatilis

The top row shows the three basic patterns; sector pattern on left, cross-banding in middle, and drop pattern on right. Below are two shells illustrating the sudden switching from one pattern to another; on left from cross-banding to drop following a change of temperature; on the right from drop to cross-banding following a change in salt content of water. (After Neumann, 1959a, 1959b)

can occur separately, or any two, or even all three can occur simultaneously. No full genetic analysis of the situation has yet been made, but it is known that some strains never exhibit the cross-banding, while another strain shows only the drop pattern. It is, therefore, probable that the realization of the pattern is controlled by a rather simple hereditary mechanism, possibly involving only one or two major genes.

What has been discovered is that a switch from one pattern to another can, in some cases at least, be brought about quite easily by simple alterations in the environment. If the sector pattern is exhibited in an individual, no changes of conditions are known which affect it. However, the drop and cross-banded patterns can be switched one into the other by changes in temperature, pH, or salt content of the water, or after a period of growth inhibition. This change, of course, only takes place if the factors for both patterns are present, and does not occur in the strain which lacks the cross-banded pattern.

The situation is perhaps rather like that of the Paramecium antigens (Beale, 1954) in that a slight alteration in the external environment brings to expression one or another of the hereditary potentialities of the individual. In *Theodoxus fluviatilis,* however, these potentialities are not strictly alternative but can both be realized at the same time, so that the cross-banded and drop patterns are superimposed. Moreover, the phenotypic characters involved are much more complex than the antigens produced in Paramecium. The formation of the cross-banded or drop patterns must both involve the interaction of quite large numbers of cellular processes, and thus necessitate the activity of many different genes. The factors, whether genetic or environmental, which cause one or other pattern to be realized must, in this case, be acting as switches between alternative systems each of which is considerably more complex than the system involved in the production of one Paramecium antigen. Perhaps a closer parallel would be the switching between the flagellate and amoeboid forms of Naegleria illustrated in Figure 39.

Area or Volume Patterns

The color patterns of the Theodoxus shell have been considered as all-over patterns because of their repetitive character. They remind one of a wall paper in which the same design is repeated many times. We now need to consider the formation of what corresponds to the individual designs in a wall paper. One of the major bodies of work on this topic is that of Henke and his students (Henke, 1933, 1947, 1948) on the color patterns of Lepidopteran wings. This well known work dates from some years ago, and I have already summarized a good deal of it in another work (Waddington, 1956) and shall therefore consider it only briefly here.

One of the first points to make, perhaps, is that in Lepidopteran wings as in Theodoxus, we are often confronted with the superposition of several different patterns, which can be recognized as physiologically distinct from one another, either because they are formed at different times during development, or because they react differentially to experimental treatments. For instance, in the wings of the moth *Plodia interpunctella*

(Schwartz, 1953) a number of different pattern systems have been distinguished which are illustrated in Figure 56. This superposition of different systems leads to increasing complexity, but is in itself not of very profound significance. The most important question is, How is any individual pattern generated?

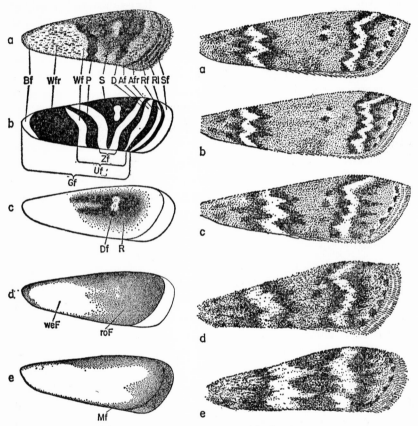

Figure 56. Patterns in wings of Plodia interpunctella

On the left, the upper drawing shows the actual pattern and the four lower drawings the various component fields into which the pattern can be analysed: b, the cross-band system; c, the ocellar system; d, the color system; e, the marginal system. (From Schwartz, 1953) On the right is a series of drawings illustrating the spread of the cross-band system in the forewing of *Ephestia kühniella* from the central region. The normal adult is at a; b, shows the broadening produced in homozygotes for Syb; c, shows narrowing in heterozygotes for the recessive lethal Sy; d and e, results of heat treatment in wild moths and in Sy heterozygotes. (From Kühn, 1955)

Henke has shown that most of the color patterns in Lepidopteran wings are brought into being gradually during the course of an on-going process. They are in fact diachronic condition-generated forms, in the sense we described in Chapter 3. A classic case is the central band which runs across the forewing of the moth *Ephestia kühniella* from front to back. This is produced by two streams of something, which begin at points on the middle of the anterior and posterior margins of the upper surface of the wing and spread from there toward the center where they meet and eventually spread sideways. If small wounds are made in the developing wing with a cautery during the first two days of pupal life, the streams of material flow round the wounded areas; while if similar experiments are made at a slightly later stage when the streams are actually proceeding, the spreading movements come to a stop so that one can get a series of adult wings that have been, as it were, frozen in successive stages in the process of pattern formation (Figure 56). A series of genes are also known which speed up or slow down the spreading process (or allow it to proceed for a longer or shorter time), so that the central band attains a greater or smaller width. Similar changes can also be produced by environmental treatments such as raising of the temperature. In wings in which there are several systems of streams of this kind, many degrees of pattern complexity can be attained by reactions occurring between two streams at the places where they come in contact.

The diachronic elaboration of patterns is undoubtedly a very important and widespread phenomenon. For instance, Kroeger (1960) has pointed out that there is a good deal of evidence that it is involved in the formation of a great many bilaterally symmetrical structures. If the development of such structures is inhibited, it is common to find that the central axial regions drop out, leaving the more lateral parts unchanged. Kroeger quotes examples based on the complex bilaterally symmetrical wing-bases of Ephestia, the genital apparatus of Drosophila, cyclopia in insects and mammals and several others (Figure 57). A still simpler example of the same type of phenomenon is seen in patterns which are symmetrical around a point rather than a line, that is to say, in radially symmetrical patterns, such as that of the ocellar spots on the wings of certain Lepidoptera. As Henke (1933) showed, increasing grades of inhibition usually result in the disappearance of the pattern from the center outward, the most peripheral ring being the last to be affected.

It is simple to explain such phenomena by the hypothesis that something
—it is difficult to say just what, the Germans can use impressive but
uninformative words like *Determinationswelle*—spreads out from the
center or axis of symmetry, and that the inhibition of development
amounts to interrupting the spread at too early a stage, or causing it to
proceed more slowly for the normal time. In either case it would be the

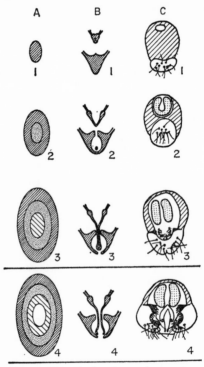

Figure 57. Reduction of patterns

Three examples of centrifugal reduction of patterns. The normal adult condition
is at 4; increasing severity of reduction in 3, 2, and 1. A is the wing ocellus of
Lepidoptera, B the axillaris-complex in the wing-base of Ephestia, and C the external genitalia in Drosophila. (After Kroeger, 1960)

central region which would be incomplete when the process comes to an
end. (An alternative explanation, which leads to the same result, is that
there is not any actual substance which spreads from place to place, but
that all regions go through a certain course of changes, the central ones
starting earlier or going faster.)

Kroeger seems to argue that this centrifugal pattern formation is a general rule; and he also suggests that the elaboration of a pattern can be adequately accounted for by supposing that, after the determining influence has spread for some distance on either side of the axis of symmetry, new centers of spread and of symmetry, come into being (Figure 58). But this, in my opinion, is to press the argument too far. In the first place, not all patterns are centrifugal in this sense. For instance, if the development of a limb-bud is inhibited, it is not the central digits which are the first to be suppressed, but rather the lateral ones; the order of effect of unspecific inhibiting agents is 1, 2, 5, with 3 and 4 the most persistent (Tschumi, 1953). The pentadactyl pattern is more nearly centripetal than centrifugal. But it is, probably, mistaken to consider it either one or the other, since with sufficiently sophisticated inhibiting agents

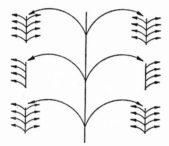

Figure 58. Secondary elaboration of pattern

A primary determination stream spreads from an axis of bilateral symmetry, and then secondary centers appear on each side of it; these secondary centers may be symmetrical, or asymmetrical. (From Kroeger, 1960)

(particular genes) many different patterns of sensitivity can be elicited; and any one of the toes may turn out to be the first to be affected. The pattern, in fact, is probably not produced by any sort of spreading process, but arises in some synchronic manner, which we shall soon discuss. And this leads to the second point concerning Kroeger's thesis; the idea that new centers of symmetry arise on either side of the original axis may be an appropriate description, but it provides no explanation of how these lateral centers come to assume independence.

The essential problem of pattern formation, in fact, is not to be found among diachronically produced patterns, since these must always be elaborations on one or a small number of underlying synchronically generated arrangements. In order to penetrate deeper into the matter, we need, therefore, to consider patterns which appear to arise synchronically.

An example may be found in some studies on the pattern of the tarsal segmentation in the legs of Drosophila (Waddington, 1942, 1948, 1953). Although much of this work was done as long ago as Henke's, it has not been very widely discussed since; but I do not know of any other example where we have as much information relevant to the problem of the factors involved in synchronic pattern generation.

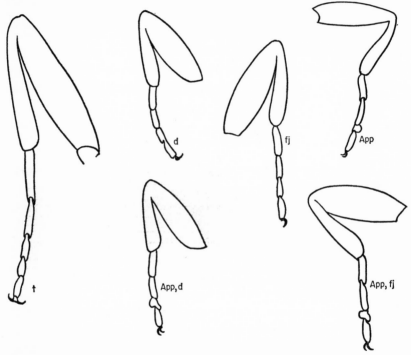

Figure 59. Tarsal segmentation in wild-type and mutants of Drosophila melanogaster *causing* four jointed *legs* (From Waddington, 1940c)

The legs of normal wild-type *Drosophila melanogaster* have five tarsal segments of characteristic length and shape. These show relatively little variation in normal populations. However, many mutant genes are known which affect the pattern of segmentation. These can be considered as falling into two main classes: those which change the normal pattern into some other almost equally definite and invariant pattern; and those which disrupt pattern formation to a greater or lesser degree, producing variable shapes which cannot be described in any simple fashion. The best known genes of the former type are *four jointed, dachs,* and *approx-imated.* In normal genetic backgrounds and environmental conditions, all these produce tarsi with four segments. The lengths of these segments

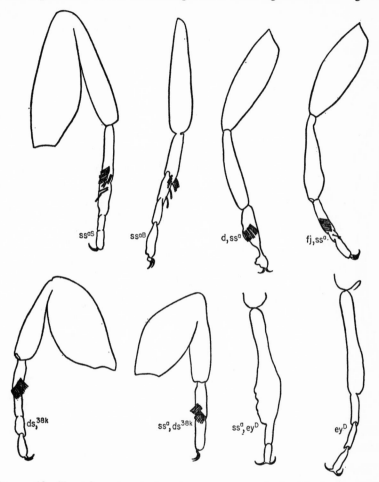

Figure 60. Tarsal segmentation in mutants and double homozygotes which disrupt pattern (From Waddington, 1940c)

show slightly more variation within a population than do those of the wild-type legs, but the difference is not very great. There are also some differences between the types of legs produced by the different genes, for instance, in *approximated* the penultimate segment is usually short and characteristically swollen. When any two of the genes are combined in double homozygotes (e.g., *dd, fjfj*), the tarsi also have four segments and there is no sign of any exaggeration of the effects of the homozygotes for single genes (Figure 59).

Some genes which destabilize the pattern forming system are *aristopedia, Eyeless-dominant, dachsous,* and *combgap* (Figure 60). Each of these genes again has a more or less characteristic effect. For instance,

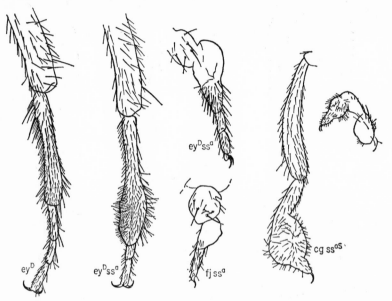

Figure 61. Local specificity in effect of segmentation genes

Eyeless-dominant produces a bottle brush effect at the proximal end of the tarsus. This is exaggerated in the legs of the *Eyeless-aristopedia* double homozygote; and the effect also occurs in the antennal leg of this genotype. Similarly the peculiar distortion of the tarsus found in *combgap-aristopedia* legs also occurs in the antennal legs.

Eyeless-dominant produces a rather recognizable "bottle brush" swelling on the proximal end of the tarsus, while *aristopedia* tends to affect the more distal regions. It is noteworthy that these effects are produced on developing leg material where ever it may be found. In particular, they can be seen in the leg which is developed from the aristal antennal bud under the influence of *aristopedia* (Figure 61). When these disruptive genes are combined in double homozygotes with genes of the four-segment group just described, there is a considerable exaggeration

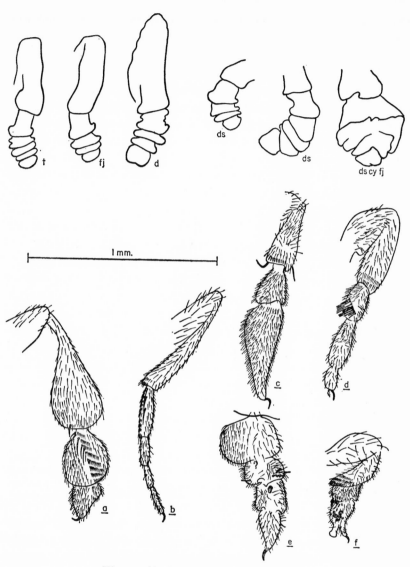

Figure 62. Disrupted segmentation

At the top is the segmentation of the imaginal-bud of the anterior leg at the time
of pupation in various mutants. At the bottom are adult legs with quite disorderly
segmentation; a and b, two hind legs from the same *combgap* individual; c, d,
second and first legs from a *dachsous combgap* male; e, f, two first legs from a
combgap four-jointed male. (After Waddington, 1943)

of effect; while in flies homozygous simultaneously for two or more destabilizing genes, very bizarre and abnormal legs may be formed in which all trace of the segmentation pattern has disappeared (Figure 62).

From these facts, one has to conclude that the normal segmentation pattern of the tarsi involves the operation of at least the seven loci which have been named above; and certainly there are many other genes, which have not been so fully studied, also active in the system. In the development of the normal wild-type leg, the activities of all these genes must mutually influence one another so as to define a fairly well stabilized creode. Another creode, not quite so stable as the five-segmented one, leads to the formation of a four-segmented tarsus and comes into operation in flies homozygous for one or more mutant alleles at the loci of *four jointed, dachs* or *approximated*. Mutations at the other loci disrupt the system of feedback interactions by which the creode is stabilized, and thus lead to the production of variable and indefinite patterns of segmentation. We could represent the situation by saying that the formation of the normal pattern depends on a balanced interaction between a number of factors, A, B, C, D, E, F, etc. Alterations in factor A produce the four-segment creode, and such alterations, of slightly different kinds, are brought about by mutant alleles of *four jointed, dachs,* and *approximated*. Changes in any of the other factors destabilize the creode. Since the four-segment creode may be supposed to be slightly less well stabilized than the normal one, double homozygotes in which both A and some other factor are abnormal will show an exaggerated effect; while if two of the factors other than A are both altered, the segmentation process will be rendered completely abortive (Figure 63).

What sort of thing should we suppose these hypothetical factors to be? The developmental processes leading up to the appearance of the tarsal segments can be divided into three main phases. There is a period in which the various imaginal buds are folded off from the embryonic ectoderm. This process has never been followed in detail in Drosophila, but probably occurs about six hours after fertilization. Then follows a relatively long period of general growth of the imaginal rudiments. This is particularly intense in the third larval instar, and toward the end of that instar shortly before puparium formation, parts of the epithelia of each imaginal bud become folded into the structures from which the adult organs will be developed. The tarsal segments first become recognizable at this period as folds in the imaginal buds. There is a good deal of evidence (Waddington, 1942) that this folding process is of fundamental importance not only for the structure of the adult organs but also for their character in the sense that abnormal folding may lead to the development from an imaginal bud of an adult structure which is normally formed by some other bud. We shall return to this later. In the present context the important point is that the pattern of tarsal segmentation is fore-shadowed by the folding of the leg-forming epithelium in the imaginal buds (Figure 62). The simplest hypothesis is to suppose that it is this folding which actually causes the segmentation. In that case, the hypothetical factors for which we are searching, are simply those conditions which control the way in which the epithelium becomes folded.

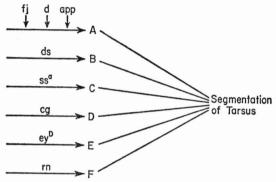

Figure 63. Effect of various genes on tarsal segmentation

The most important variables in this connection would be the over-all size of the epithelium and the various conditions of surface tension, viscosity, elasticity, and so on characterizing the epithelium as a whole. We can imagine the leg epithelium forming a series of folds, as a drying blob of paint forms a series of wrinkles on its surface. There are, presumably, first a set of major folds which define the tarsus, the tibia, femur, and so on; and then a set of secondary folds by which the tarsus itself becomes divided into segments. The distance between each of the folds would be dependent partly on the total mass of material available, and partly on the properties of the material in relation to the physical factors just mentioned. Now, the viscosity, elasticity, surface tension, etc. of an epithelium must clearly be the resultant of its particular molecular composition, and could be altered by the production within it of different enzymes, etc. Genes of the normal kind might be expected to alter these properties, and thus affect the final pattern in which the tarsus becomes segmented. There seems to be no need to suppose that pattern formation depends on a special class of pattern-determining genes, or higher fields of genetic action of the kind to which Pontecorvo (1959) has referred.

I think the same conclusion stands even if we adopt a less oversimplified theory of the way in which the segmentation pattern arises. It is probably true that such crude variables as tissue elasticity, viscosity, etc. are only part of the story. When one examines the legs in more detail, one finds such things as the characteristic swelling of the penultimate segment in *approximated* (Figure 59), the localized bottle brush in *Eyeless-dominant* (Figure 61), and various other evidences of a local specificity in action which can scarcely be accounted for except by supposing that several species of very complex molecules must be involved. But nothing forces us to introduce anything more into our armory of genetical concepts than we have to invoke to explain the relation between genes and macromolecules in general.

All the tarsus segmentation genes just described are pleiotropic and have easily recognizable effects on other parts of the body. In the three genes producing four-segmented tarsi (*dachs, four jointed,* and *approximated*) these effects are comparatively slight. They amount to the production of a shorter, squarer wing, in which the longitudinal veins diverge at a slightly greater angle than usual, and as a consequence of this the two cross-veins are closer together. The other pattern-disrupting genes have more drastic effects. *Eyeless* reduces the size of the eyes, *aristopedia* converts the arista into a leg, *combgap* produces absences of venation and often quite abnormal wings, and *dachsous* makes the whole exterior structure of the adult, including wings, body and eyes, abnormal. In double homozygotes for these genes, the effects on other body regions are often very exaggerated, sometimes even more so than the effects on the legs themselves. The effects are often of a kind which must have required alterations of growth rate during the larval period. For instance, in *dachsous-combgap,* the thorax is sometimes obviously too large, and swells up into large shoulders on either side (Figure 64). In these double homozygotes, we may also find cases of abnormal histogenesis, for instance, the conversion of a wing into a part of thoracic surface, or of the eyes into a palp or even into an antenna. Some organs may also be duplicated. Perhaps the prize is taken by a *combgap-four jointed* fly, whose head carried one very reduced eye and *five* antennae. These abnormalities can probably all be explained as the results of abnormal foldings of imaginal tissue at the end of larval life, resulting from changes in the growth rate and physical properties of the epithelia from which the adult organs will develop.

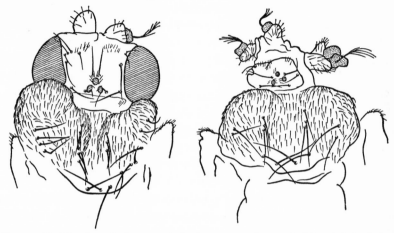

Figure 64. Head and thorax of dachsous combgap *(left) and* dachsous combgap-four jointed *(right) fly* (From Waddington, 1943)

Spot Patterns

The most elementary type of spot pattern consists of a scatter of small areas of different sizes randomly distributed over an expanse. Such a "fleck-pattern" as Henke (1947) called it, could arise by simple stochastic variation of the kind considered by Turing, which I discussed in Chapter 3. With this as the initial stage, we can imagine patterns of greater regularity to arise in two different ways. If the individual areas are increased in size and run together into groups, the pattern will develop toward the various types of area pattern which Henke discussed and which we mentioned earlier. In this section we are concerned with the other possible form of regularity: that in which the spots remain small but their distribution becomes more orderly.

The most obvious mechanism by which this can come about is for the spots to interfere with one another in some way. Wigglesworth (1940) has used a hypothesis of this kind to explain the pattern of hairs on the larval abdomen of the bug *Rhodnius prolixus*. The external surface of the abdomen is covered with a large number of small hairs which are evenly spaced and approximately equal distances apart. Wigglesworth suggests that this even spacing is due to the fact that each hair attracts to itself some necessary precursor within the hypodermis, and therefore prevents the formation of any other hair within a certain radius. This type of competitive action between hairs would indeed account for the even distribution, provided (a point not specifically emphasized by Wigglesworth) that as each hair reached its normal final size, its competitive efficiency falls off. Unless there is some self-limiting action of this kind, one would expect that those hairs which began by chance to have an advantage over their neighbors, would "run away," and become much too large with an exaggerated space around them.

The most thorough study of competitive interactions between locations in a spot pattern is concerned with the development of hair follicles in the skins of sheep. The enormous economic importance of the wool crop has stimulated a rather thorough study of this system. An interpretation of it in terms of competition between follicles for a precursor was first formulated by Fraser (1952) in our laboratory in Edinburgh, and it has since been developed by him, and a number of others in Australia. A recent summary is Fraser and Short (1960). The whole system that has been disclosed is somewhat complex, and all that it is possible to do here is to give an indication of the general nature of the variables involved, and the way in which they are related.

The hair follicles in the skins of sheep fall into groups, each group consisting of a number of primary follicles (usually three) with which are associated a cluster of secondary follicles (Figure 65). The number of secondaries per primary is an important variable. It is about four in the relatively primitive carpet-wool breeds, and between five and six in the long-wool and down breeds which have been developed from them. In the Merino, which is the basis of the Australian wool industry, the secondary to primary ratio may be as high as twenty. Large differences in secondary to primary ratios, such as that between the Merino and the other breeds, depend rather directly on genetic changes; and it is usually considered that the Merino breeds were independently evolved from wild sheep, and are not derivatives of any of the other improved breeds

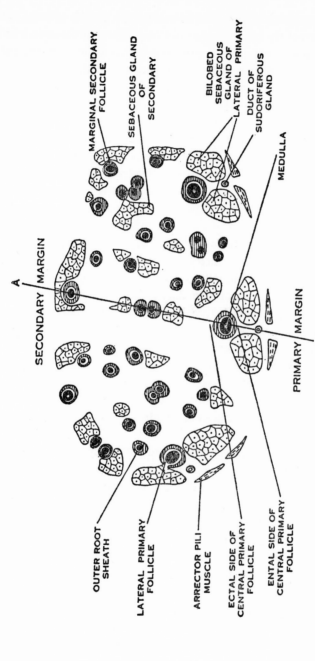

Figure 65. Group of wool follicles in transverse section (From Fraser and Short after Hardy and Lyne, 1960)

SECONDARY MARGIN

PRIMARY MARGIN

A

B

MARGINAL SECONDARY FOLLICLE

SEBACEOUS GLAND OF SECONDARY

BILOBED SEBACEOUS GLAND OF LATERAL PRIMARY

DUCT OF SUDORIFEROUS GLAND

MEDULLA

OUTER ROOT SHEATH

LATERAL PRIMARY FOLLICLE

ARRECTOR PILI MUSCLE

ECTAL SIDE OF CENTRAL PRIMARY FOLLICLE

ENTAL SIDE OF CENTRAL PRIMARY FOLLICLE

in which the gene for high secondary to primary ratio is lacking. Within any particular breed the actual number of secondaries associated with a particular group of primaries is probably influenced by competition between the follicles during their development. The thing competed for can be called in very general terms "follicle substrate," but is perhaps better referred to as "development space." Although this term is also intended to be understood in a rather abstract sense, real physical room for the follicles to develop is, probably, one of the main components for which competition occurs, at least in Merinos, in which the large number of secondaries are formed partly by budding from one another and form a compact mass of tubules between which there is little room.

The competition between follicles is complicated by the fact that they are initiated at different stages of development and grow at different rates. The major phases of development of a follicle group are four in number. First, the central primary follicle is formed; second, the lateral primary follicles appear, usually one on each side of the center; third, the formation of a number of secondary follicles which mature and begin to form keratinized fibers before birth; finally, another group of secondary follicles appear, which do not begin to secrete keratinized fibers until after birth. The number of follicles between which competition can take place, therefore, increases as time goes on. Further, the competitive efficiency of the different follicles does not remain constant as they grow and develop. All these variables, of time of initiation, rate of maturation, and change in competitive efficiency, can be separately affected by genetic control.

This gives rise to a highly complex and flexible system. The operation of this has not been fully studied in relation to follicle formation, but has been investigated in greater detail in connection with the formation of hairs by those follicles which succeed in entering into production. It is a fortunate circumstance that the direction of elongation of a wool fiber changes periodically during its growth, being turned through an angle of about 180° during a period of about a week, after which it turns back in the opposite direction again. This gives rise to the phenomenon known as the "crimp," that is to say, the waviness of a natural lock of wool as it is clipped from the sheep. The crimp period can be used as a clock to time the rate of growth of various fibers derived from different types of follicle; and on this basis we arrive at some evaluation of the various factors involved in the competitive system which has been described.

The spatial distribution of the wool follicles over an area of skin shows only a moderate degree of orderliness (Figure 65). No general theory exists of just how much order the processes of competition would be expected to engender. Many of the spot patterns we come across in biological materials are rather highly ordered, and it seems likely that they usually involve something more than mere competition of the kind we have been discussing. For instance, on the thorax of Drosophila, there are two types of hair, the macrochaetae and the microchaetae. The latter, which probably correspond more or less to the hairs on the abdomen of Rhodnius discussed by Wigglesworth, are arranged in orderly anteroposterior rows. The adult thorax is, of course, produced by the fusion of two half thoraxes, one developed from each of the two mesothoracic imaginal buds. If one of these buds fails to become everted on to the surface, only half a thorax appears, and this is often quite distorted in general shape. Nevertheless, the microchaetae still occur in definite rows which are unrelated to the general configuration of the thoracic surface. In some flies of this kind, one mesothoracic bud forms considerably more than a half thorax, and in this extra region also the microchaetae lie in neat anteroposterior order (Waddington, 1953). It is clear, therefore, that the formation of this pattern involves something more than mere competition between the hairs. The same conclusion is very obviously true of the pattern of macrochaetae, which are formed in positions the distances between which are neither equal nor related to the size of the bristles.

Strong evidence that the pattern of bristles on the Drosophila thorax is based on a system of specific local differences, rather than on any general system of competition extending over the whole area, can be found in the fact that it is possible to produce animals in which one part of the pattern has been removed without any noticeable effect on the other neighboring parts. A well-known phenotype in Drosophila is "bithorax," in which the metathoracic buds are persuaded to develop into adult thoracic parts similar to those usually formed from the mesothoracic buds. This condition can be brought about in a number of different ways; by a series of alleles of the *bithorax* gene which we shall discuss in a moment, by extreme environmental stresses such as exposure to ether vapor applied to the early developmental stages of the egg, and by a strain in which such environmentally induced modifications have been genetically assimilated following a process of selection (Waddington, 1957). In this assimilated bithorax strain the metathorax is often of

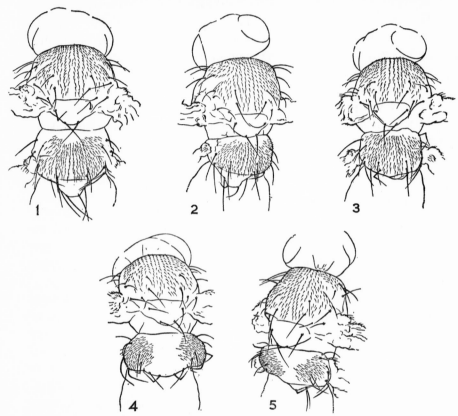

Figure 66. Hair patterns on metathoracic mesothorax in flies of assimilated bithorax stock

In 1 is a complete representation of the normal pattern. In 2, some of the anterior bristles (presuturals and notopleurals) are missing, while in 3 the posterior ones (scutellars) are absent. In 4 and 5 are two examples of complete half-patterns which have not fused properly in the mid-dorsal line.

considerable size. When most fully developed, it may exhibit the complete bristle pattern of the mesothoracic region (Figure 66), but in some individuals the anterior part of the pattern may be missing, in others, the posterior part. In both cases of partial reduction, the region which is formed seems to be quite normal and uneffected by the absence of the missing section. The pattern in fact behaves in a mosaic manner, as though each part is independent of the other parts. (Actually one would, I think, expect to find that the complete pattern consisted of a mosaic of subpatterns within which mutual interaction of parts could go on. However, owing to the comparatively small number of macrochaetae involved, the situation does not easily lend itself to such detailed analysis.) A similar independence of the separate parts is, of course, found between the two lateral halves of the thorax, which are derived from different imaginal buds.

The development of the bithorax phenotype provides a favorable example for discussing the general problem of whether there is a relation between the spatial parts of the adult phenotype and the spatially ordered regions of the chromosome. A possible hint of such a relation can be found in some extremely interesting results obtained by Lewis (1955), who studied a number of pseudoalleles of the *bithorax* locus in the third chromosome of Drosophila. He identified five distinct mutant alleles concerned with transformations of the meso- and metathoracic imaginal material. Although these transformations are not directly changes in a spot pattern, but are alterations of an area from a metathoracic to a mesothoracic character, this is quite a convenient place to discuss them.

By crossing-over analysis, it was shown that the five alleles lie along the chromosome in the order *bithorax, contrabithorax, ultrabithorax, bithoraxoid,* and *postbithorax.* Each allele effects certain particular regions and changes its character in a particular way. The question is, can any relation be demonstrated between the order of the alleles on the chromosome and the geographical arrangement of the regions which they affect? The following list shows the order of the genes and the type of transformation which they bring about:

Bithorax   anterior metathorax into anterior mesothorax.
Contrabithorax   posterior mesothorax into posterior metathorax.
Ultrabithorax   anterior metathorax into anterior mesothorax.
Bithoraxoid   posterior metathorax to mesothorax, and anterior first abdominal to anterior metathorax.
Postbithorax   posterior metathorax to posterior mesothorax.

It seems to emerge fairly clearly that there is in fact no direct connection between chromosomal order and the order of the phenotypic effect on

the anteroposterior axis. For instance, *bithoraxoid* affects a more posterior region (the posterior metathorax and the anterior abdomen) than
either *bithorax* or *postbithorax;* but it lies in between these two alleles.
It seems impossible, in fact, to derive from this situation any plausible
scheme of a higher field of organization of the genetic material (of the
kind that Pontecorvo [1959] alluded to as a possibility) which would
have a structure in any way reflecting that of the adult phenotype.

Lewis has studied the position effects following on chromosome break-
age and rejoining in this region, and the cis-trans effects found in various
heterozygotes for these different alleles. He interprets his results in terms
of the sequential transformation of a substance which passes along the
chromosome region from one end to the other. This is, of course, one of
the recognized forms of explanation of phenomena of this kind, and I do
not wish at this point to get involved in the controversy, interesting
though it is, as to the mechanism of position and cis-trans effects. The
point at issue in the present context, is that there are no grounds for
stating that, for instance, the substance produced in this region becomes
capable of acting on more and more anterior regions as it passes from
one end of the set of pseudoalleles to the other, or that it follows any
other simple rule of correspondence of this kind.

Geneticists who wish to explain everything without lifting their eyes
from the chromosomes—if there are any such—may however challenge
the embryologist to show what type of explanation he would expect to
find applicable to a situation such as that of the *bithorax* alleles. I shall,
therefore, put forward a completely speculative model, simply to serve
as an example of one of the possible kinds of thing which might be going on. The character of an imaginal bud, as either mesothoracic or
metathoracic, is probably strongly influenced, though possibly not finally
determined, at the period in which it is first separated from the general
ectoderm of the embryo; this is suggested by the fact that phenocopying
actions can be exerted at this period, for example by exposure to ether
vapor. Suppose now that the region of ectoderm which will be folded
off to form the imaginal buds decreases in width from anterior to posterior, and that the segment quality (mesothoracic or metathoracic)
which it eventually shows depends on the width of material at the time it
is folded into the bud. Some support for such a suggestion can be found
in the fact that we have reason to believe that abnormal folding of the

imaginal buds at a later stage (i.e., at the end of the larval period before they enter the pupal stage of development) may cause tissue to be switched from, for instance, eye to antenna or antenna to leg (Waddington, 1942c). We could then account for the effect of the various *bithorax* pseudoalleles in terms of their influence on the growth of the ectodermal rudiment up to the stage at which it becomes segregated into definite imaginal buds. A possible scheme is indicated in Figure 67. The various

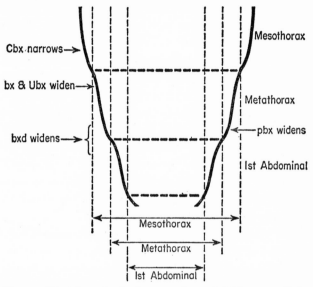

Figure 67. Hypothetical operation of bithorax alleles

It is supposed that the embryonic ectoderm from which the imaginal-buds will be folded off has the shape shown, in dorso-ventral view, by the heavy lines. This material is supposed to develop into mesothorax if it is wider than a certain limit; into metathorax if it is somewhat narrower; and into the first abdominal segment if still narrower. The width limits for these segments are indicated below. Alteration of width may change segment quality, but does not change an anterior part of a segment into a posterior part (cf. studies of Locke, p. 200). The phenotypes produced by the various alleles would then be explained if they controlled the growth of the embryonic rudiment in the manner indicated.

alleles would then act by the production of some characteristically modified enzyme which controls the width of the imaginal bud rudiments at particular places. This does not involve anything more surprising or difficult to comprehend than, for instance, the suggestion that there are genes for a Grecian, a Jewish, or a Negro nose, which we accept without question. A scheme of this character has some general biological plausibility, and it suffices to show that the facts of the *bithorax* situation, odd and challenging though they are, do not exhibit any features which are difficult to incorporate into an orthodox embryological theory.

Returning to the previous discussion of mosaic-like patterns, these need not necessarily be spot patterns. For instance, an example is provided by the effects of the various alleles of *cubitus interruptus* on the venation pattern of the Drosophila wing, illustrated in Figure 68. Mutant alleles at this locus cause interruptions in the third and fourth

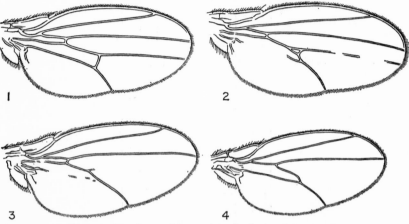

Figure 68. Smoothing out of abnormal pattern

The normal wing venation of Drosophila is shown at 1. In *cubitus interruptus,* abnormalities such as 2 and 3 are produced by most alleles, but in *ci^W* wings such as that shown in 4 are rather common.

longitudinal veins. The study of developing wings shows that the action of the more extreme alleles can be already seen in the venation at the pre-pupal stage, but further tendencies to obliteration of veins continue well into the following pupal instar. When only comparatively slight deviations from the normal phenotype are produced, the action is usually of a mosaic kind with the wing looking as though it carried the normal pattern of veins except that certain portions have been wiped out. When the abnormality is more extreme, however, one quite commonly finds a pattern, such as that illustrated in Figure 68, in which a new integration of portions of the fourth and fifth longitudinal vein and of the posterior cross-vein has taken place, so as to produce a pattern with few loose ends, or isolated segments of vein. The simplest hypothesis, of course, is to suppose that the mosaic type of action, which hits certain parts of what looks like an initially complete venation pattern, occurs when the mutant gene becomes effective relatively late in development, while if it acts at an earlier stage, the altered elements in the pattern-forming system can react with one another to produce a new organized pattern.

It is worth emphasizing that pattern-forming systems in biological materials seem frequently to be able to produce relatively well organized patterns even when they have suffered alteration, for instance by the presence of mutant genes. I have referred to this on an earlier occasion (Waddington, 1957), where I suggested that during evolution there might be a form of "archetypal" selection for genes which modify pattern-forming systems in such a way that the production of integrated patterns remains possible. We have seen some examples of such genes, not only in the *cubitus interruptus* allele illustrated in Figure 68, but in the genes *dachs, approximated,* and *four jointed,* which produce abnormal but regular patterns of segmentation of the tarsus (Figure 59). Other evidence of the tendency for the production of integrated patterns—and in this case something which can almost be considered as a spot pattern—can be found in the ommatidia of Drosophila eyes affected by various mutants. In many such mutants the individual ommatidia may contain some number of fully formed rhabdomeres other than the usual seven found in the wild-type. In some cases all eight retinula cells produce a rhabdomere; in other cases the number of rhabdomeres is less than seven and may be highly variable from one ommatidium to the next. In eyes in which the number of rhabdomeres varies considerably from ommatidium to ommatidium the pattern they form in cross-section is, of course, also very variable. Nevertheless, it is rather surprising to find how often two or more neighboring ommatidia show almost identical patterns. Some examples of this are illustrated in Plates XXII to XXIV. The phenomenon seems only explicable on the hypothesis that the pattern is formed as the result of interactions between the retinula cells, and that there are a certain number of more or less stable patterns at one or other of which the interactions tend to settle down. The concept of "positions of organic stability" is a very old one, going back at least to Galton's *Natural Inher-*

itance published almost a century ago. The repeatability of abnormal patterns in the ommatidia of these mutant eyes is good evidence for its validity.

It is possible to carry the analysis of the stability of normal patterns a good deal further than this, since they can be obtained in large enough numbers. The most penetrating investigations so far have been by genetic methods. In wild-type populations, many patterns seem to be almost invariant. For instance, in Drosophila the scutellum normally carries four bristles, arranged in a rather definite way at the four corners of a trapezoid, and in most stocks, individuals which vary from this pattern are very rare. In strains containing certain mutant alleles more variability in the number of scutellar bristles may be encountered. This is so not only when the mutant allele is one which has a well-recognized effect on this part of the animal, such as alleles of the locus *scute,* but certain other genes, whose most important effects are elsewhere, may also cause the appearance of somewhat increased numbers of individuals with abnormal scutellar bristle patterns. One such gene, for instance, is *Eyeless-dominant.* This allele has a recessive lethal effect, so that a strain permanently segregating for *Eyeless-dominant* and wild-type can be maintained by always using heterozygotes as parents. In such a strain the *Eyeless* individuals show considerable variation in number of scutellar bristles, the wild-type individuals very little. If selection for increased number of bristles is practiced on the *Eyeless* flies, the number increases from generation to generation, showing that there is within the population some genetic variability concerned with the number of these bristles. If genes tending to cause an increase in scutellar bristles are accumulated sufficiently by selection of this kind, the wild-type sibs will eventually also come to show some increase in bristle number (Waddington, unpublished).

The fact that the wild-type individuals show much less phenotypic expression of the presence of genes tending to increase bristle number than the *Eyeless-dominant* ones do, is evidence that in them the normal four-bristle pattern is in some way stabilized or buffered. The effect of the abnormal gene, in this case *Eyeless-dominant,* must have been to produce some destabilization of the pattern, so that the inherent genetic variability could come to expression and be submitted to selection. The process is very similar to that in which an environmental stress is utilized to destabilize a developmental system and to reveal genetic variation which was previously concealed. Selection on this variation may eventually lead to the genetic assimilation of the phenotypic modification produced by the environmental stress (Waddington, 1961).

The experiments on scutellar bristles in *Eyeless-dominant* strains were carried out some years ago and without any great quantitative precision. More recently, and independently, Rendel (1959; Rendel and Sheldon, 1960) has made an accurate study of the response to selection for scutellar bristle number of strains in which this phenotypic pattern had been destabilized by the presence of an allele of *scute*. He was able to show quite conclusively that the wild-type phenotype can be produced by individuals which differ considerably in genetic factors tending to affect bristle number. The wild pattern is so well stabilized that such genetic variations do not come to expression unless they surpass either an upper or a lower threshold, which lie some considerable distance apart. If the assumption is made—and there is a good deal of evidence to justify it as at least first approximation to the truth—that the strength of genetic bristle-producing tendencies is normally distributed in each of the populations involved in the experiments, it is possible to deduce a scale by which the intensity of bristle-producing tendency can be measured. A graph can then be drawn showing the relation between the number of bristles actually produced and the strength of the genetically-determined bristle-producing tendencies (Figure 69). The stabilization of the pattern is expressed in the fact that quite a wide range of genetic intensities result in the formation of four bristles. In fact, to increase bristle numbers from three to five, required about eight times as great an increase in genetic intensity as it did to produce two extra bristles in the range from one to three.

These experiments lead us to the same general conclusion as we reached in discussing volume patterns, such as those of tarsal segmentation; namely, that the normal pattern, in spite of its apparent simplicity, is produced by the interaction of a large number of different factors, each separately under genetic control, and further, that this complex system of interacting processes reaches an end result which is stable in the sense that it is invariant even in the face of considerable variation in the intensity of the constituent processes. We are, in fact, dealing once again with development along a creode or stabilized pathway, although in this case the end state of the creode is not a certain complex chemical constitution, as it is in the development of a particular histological type of cell, such as muscle or nerve, but is instead a definite spatial distribution of the elements concerned in a pattern, such as scutellar bristles.

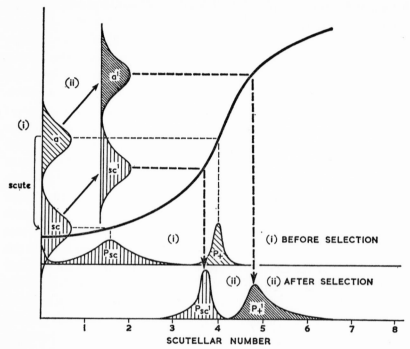

Figure 69. Results of selection for bristle number in populations segregating for scute *in Drosophila*

The heavy sigmoid line represents the relation between dosage of bristle-producing genes, represented vertically, and phenotypes, represented horizontally. In a wild-type population before selection, there is some variation of gene dosage (curve a) but this results in very little variance in phenotypes (P+). In the unselected scute sibs, the genetic variance (curve sc) is expressed by considerable variance in phenotypes (Psc). After selection for high bristle number in the scutes, the population genotypes are shifted to a^1 and sc^1, and these result in less phenotypic variance among the scutes (Psc^1) and more among the nonscutes ($P+^1$). (The relation between genotype and phenotype is represented as unchanging, but actually it would probably be altered by the selection in such a way as to reduce the canalization.) (After Rendel, 1959)

Fraser and his associates (Dun and Fraser, 1959; Fraser and Kindred, 1961) have carried the analysis of another spot pattern still further from a similar point of view. The character in question is the number of vibrissae in the mouse. These are long sensitive hairs which are located at particular regions on the mouse's head and a few other parts of the body (Figure 70). In normal mice, there is very little variation in vibrissa number, but in the presence of the sex-linked partial dominant *Tabby,* the number is considerably reduced and a good deal of genetic variation is revealed which can be subjected to selection.

It was first shown that selection for vibrissa number in *Tabby* is effective, just as is selection for scutellar bristle number in *scute* Drosophila. Again, the genetic variability utilized in this selection is present in the wild-type strains, but does not come to expression owing to the canalization of the normal developmental pathway (Figure 70).

A further refinement of analysis is possible owing to the fact that the mouse contains a number of different types of hairs in its coat. The vibrissae themselves fall into two categories, which can be referred to as primary and secondary vibrissae; and the main coat contains four main types of hairs, known as guard hairs, awls, auchenes, and zigzags. In *Tabby* mice the variability in vibrissa number is mostly concerned with the number of secondary vibrissae, and the selection which was made in Fraser's experiments was mainly based on the number of hairs of this kind. By the sixteenth generation of selection, there was a considerable difference between the high selected and the low selected lines in the number of secondary vibrissae. The crude difference in total number of secondary vibrissae is, however, a composite figure. If the hair patterns are examined in more detail, it is found that there are certain groups of hairs the number of which had scarcely been increased in the upward selection line although it was considerably lowered in the downward selection, while in other groups the number could both be somewhat reduced and also considerably increased. The results can be interpreted if we suppose: first, that there is a zone of relatively easy phenotypic modification, protected on either side by thresholds, so that it is very difficult to increase the number of bristles in a group beyond a certain number or to reduce it below some other number; second, that in the presence of *Tabby* some groups of vibrissae remain near the upper boundary of this zone, so that they can respond to downward selection but hardly at all to upward selection, while other groups are brought to a position nearer the middle of the zone, so that they can respond in both directions.

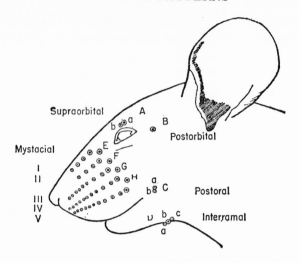

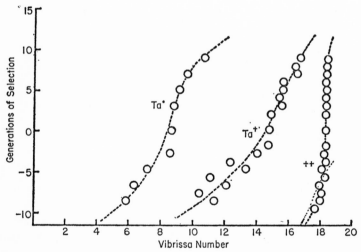

Figure 70. Vibrissae in the mouse

The positions of the main vibrissae in the mouse are shown above (there are also some on the feet). At the bottom is the effect of selecting for increased vibrissa number in the *Tabby* heterozygotes of a population segregating for this gene. The number of vibrissae in the well-canalized wild-type is increased only very slightly. (After Fraser and Kindred, 1960)

The second main point that emerged relates to differences in behavior between the secondary vibrissae in general, and the primary vibrissae and the hairs of the main coat. Regarded as a whole, the secondary vibrissae responded to selection moderately well, but this same selection produced hardly any alteration in the number of primary vibrissae. On the other hand, in the main coat the effect was rather striking. In the upward selection line the number of auchenes was increased and of awls

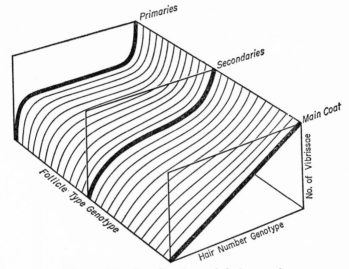

Figure 71. Canalization of hair number

The heavy lines represent the genotype-phenotype relationships for the primary vibrissae, the secondary vibrissae, and the main coat. The first is well canalized; the last, very little canalized with respect to changes in the genotype controlling hair number. There are also genes affecting the partition of follicles into the three types, so that the whole genotype-phenotype relationship requires representation as a surface, which is indicated by the fine lines. (After Fraser, in press)

reduced, while the opposite effect was seen in the low selection line. The effects were sufficiently marked for it to be possible, on the basis of the character of the main coat, to classify an individual mouse as coming from the high or low line with only a slight probability of error. This seems to imply that the canalization of the character of the main coat is rather slight, so that it exhibits quite easily the effect of genetic variation. The canalization of the number of secondary vibrissae on the other hand is moderately strong, and that of the number of primaries very strong.

Fraser provides a diagram (Figure 71) of the relations between the phenotype and the various sorts of genetic variation which can affect it. The diagram involves three dimensions. One of these is concerned with deciding whether a given follicle shall be a primary vibrissa, a secondary vibrissa, or part of the main coat (the follicle type axis). Another is concerned with the basic genotype for hair production, and *Tabby,* as well as the genes concerned in the response to the selection, bring about variations along this axis. The third axis indicates the nature of the phenotype, that is, number of vibrissae or character of main coat. The relation between these three variables is indicated by a surface. This is curved in a way which corresponds to the high canalization of the primary vibrissa number, the moderate canalization of the number of secondary vibrissae, and the lack of canalization of the character of the main coat. As we can deduce from other experiments (e.g., Waddington, 1960; Rendel, 1959) the shape of this canalization surface will itself be under genetic control, but this is not indicated in the diagram.

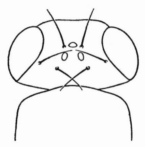

Figure 72. Location of ocelli and associated bristles in wild-type Drosophila (From Smith and Sondhi, 1960)

By analyzing patterns with genetic means, it has often been possible to show that different parts of a single pattern may be separately affected by genetic variations. One instance of this is the different behavior of the various groups of secondary vibrissae just described. Another example occurs in the experiments of Smith and Sondhi (1960) on the pattern of bristle and ocelli on the dorsal surface of the head of Drosophila. There are normally three ocelli, which lie at the corners of a triangle, with its apex pointing toward the front of the head; associated with these are three pairs of bristles (Figure 72). The complete pattern is produced by the fusion of the two half-heads, each derived from one cephalic imaginal bud; and the anterior ocellus which lies on the mid-line is brought into being by the fusion of two half rudiments (exactly how these two half-ocelli succeed so often in making a perfect fusion remains somewhat mysterious and would repay investigation). This pattern, again, shows little variation in wild-type flies, but it can be destabilized by the mutant *ocelliless*. Selection in such *ocelliless* flies shows not only that there is genetic variation affecting the total number of bristles, but also that one can shift the distribution along the antero-posterior axis, so that one obtains, for instance, a strain in which the posterior ocelli and bristles are usually present in their normal numbers while the anterior ones are absent. On the other hand, it was not possible to concentrate the pattern on to one side or the other, which would give rise to strains in which the left ocelli and bristles, or alternatively the right ocelli and bristles, were predominantly developed. This is perhaps hardly surprising, since one could scarcely expect to find genes which operate differentially as between the right and left cephalic imaginal buds. Another instance in which it has been possible to shift the balance of a pattern within a coherent and connected region of tissue, that is, within an imaginal bud, is provided by the work of Reeve (1961), who showed that selection could increase the number of the more ventral hairs in the sternopleural group in comparison with the more dorsal ones.

Conclusion

These excursions into quantitative genetics reveal to us something of the complexity of the systems of interacting components which must be involved in the development of even relatively simple phenotypic patterns. By using a genetic method of investigation we are, as it were, jumping in one step from phenotypic characters, such as hairs in particular positions, to the basic genetic (chemical) factors which can affect the developmental processes by which the phenotype is brought into being. In doing this we are presumably failing to identify any unitary components of the system which may be effective at intermediate levels of complexity. If, for instance, in a particular developmental system involved in pattern formation an important part is played by factors such as the permeability or tension of a cell surface, or the resistance to bending of an epithelium, we should not, by genetic analysis, be brought to realize this. We should instead find that very many genetic factors were active, but we should not be able to discover that many of them are operative because they affect such factors as cell membranes or elastic properties. The genetic analysis of patterns, therefore, reveals to us something about their complexity in terms of ultimate units, and also something about the degree of integration of these ultimate units into stabilized creodes, but it is much less informative about the intermediate steps between the genes and the final patterns. These intermediate steps have been the main concern of embryologists; and we saw, in the first parts of this chapter, that there are still many problems which must be tackled at the level of complex components rather than of the basic genetic elements. The fact that a certain gene-substitution in Drosophila causes the appearance of a four-segmented leg in place of a five-segmented one should be regarded as facing us with a problem and not as providing us with an explanation. The task of biologists who explore the genesis of pattern is to identify the immediate causes for the break-up of a uniform region into separate elements situated in a definite spatial order. And I drew attention, in the earlier part of this chapter, to a rather unexpected fact which I suggested may be very relevant to this problem; namely, that similar-appearing items in a pattern may actually each have individual characters which are specifically different from each other.

Epilogue

A *S* I gave warning at the beginning of this book, the discussions in the preceding chapters have been concerned not with one but with two themes. I have not, of course, attempted within the confines of a single, fairly small, book to review either of these exhaustively. I have limited myself to mentioning a number of points that seem to me particularly interesting and significant in relation, on the one hand, to the development of new lines of thought concerning developmental processes; and on the other, to the problem of how the structural organization of biological materials is brought into being. In both these contexts, I have tried to present the discussion in a way which allows us to keep at least one foot on the solid ground of fact. Even the most far-reaching and illuminating of the new patterns of thought have taken their origin, or at least have derived their strength, from a foundation of new experimental observations. I shall now make some concluding remarks about the kinds of new fact which we have recently been obtaining and which we may, with reasonable prospect of success, set ourselves to discover in the near future.

I think few people will deny that the major source of new ideas concerning development has, in the last two decades, been provided by the series of brilliant discoveries about the control of protein synthesis by genes. Many of these discoveries have been made by geneticists studying microorganisms. Another large group of them have come from the investigation by biochemical and biophysical methods of relatively complex macromolecules. The achievement of these two types of work have been so magnificent and undeniable that one feels a considerable temptation to conclude that all the fundamental problems of biology will yield to attack by the methods of molecular biology and microbial genetics; or at least that this is overwhelmingly the most favorable route of advance for the near future. Molecular biology is the fashion; and I for one am happy to see it in that position. But just how exclusive a fashion should it be?

I think one's attitude to a fashion in science should always be ambivalent. First, one should try to carry it along its own lines as far as it will go, but second, one should try to discover the irritant facts with which it is failing to deal and which may succeed in switching it eventually into a new path. Perhaps I can claim some justification for taking this two-faced attitude from my experience with the previous fashion in developmental biology. When I started my research career, the fashion in this subject had been set by Spemann's recent discovery of the phenomenon of embryonic induction. I began by pushing straightforward along the "organizer" line and showed that Spemann's principles, which had been founded exclusively on experiments with amphibian embryos, could also be applied, with suitable modification, to the quite different embryos of birds. But after that I felt the need to bring in points of view which lay outside the current fashion; and the two points of view which then seemed necessary but unfashionable were, in fact, those which have become fashionable now. One of these was biochemistry, which at first took the form of attempts to show that the induction process was a chemical one and to identify the nature of the chemical substances involved. The second was genetics, the thesis being that the synthetic processes by which development is brought about are essentially based on genic action. By the end of the 1930s these two lines of thought, the biochemical and the genetical, had come together into the concept that the genetic material consists essentially of a linear array of relatively small molecular groupings (see for instance the last few pages of Waddington, 1939). Although at that time it was considered that these groupings were likely to be amino acids and not, as we now think, nucleotides, this idea was one of the rudimentary primordia from which recent molecular biology has grown. If then I proceed for the next few paragraphs to act as devil's advocate against the present fashion, I feel I cannot be accused of being wholly unsympathetic to it. The main point I want to urge is that just as the fashionable embryology of the 1930s needed to be amplified by the addition of some chemistry and genetics, so the presently fashionable biochemical genetics of microorganisms needs to be enriched by the addition of some metazoan embryology.

It will be best to consider these matters in connection with specific and definite problems, but before we do so it is necessary to say a few words about what we mean nowadays by genetics or embryology. Genetics began as the study of the transmission of characters from one generation of multicellular organism to its offspring in the next generation. It was, therefore, concerned with simple but stable units which determined potentialities which gradually became unfolded during development to bring into being the characters in question. More recently methods have been found for studying the transmission of characters from one generation to another of a single-celled organism. In such cases, there may be transmission not only of simple units representing potentialities, but also of fully-developed components of the organism, such as for instance, mitochondria, chloroplasts, and the like. The distinction between the genotype as that which is passed on in heredity and the phenotype as that which is developed on the basis of the transmitted potentialities, becomes difficult to draw. Finally, the word genetics is nowadays often used in connection with the behavior in time of the characteristics of cells taken from a multicellular organism. Insofar as this behavior involves persistence from cell generation to generation, or changes which are brought about by processes of cell division and fusion, we are still dealing with processes which are closely similar to those involved in genetics in its original sense. However, it is almost impossible to draw a sharp dividing line between these and other changes which take place in the characters of cells without any relation to cell division; for instance, changes which may occur in the period between two cell divisions or after cell division has ceased. These changes unrelated to cell division are, of course, brought about by the working out of the potentialities inherent in the cell. They are, in fact, developmental changes and certainly belong to the field of embryology. There is therefore no strict dividing line in the conceptual realm between genetics and embryology; in such subjects as cell genetics, they merge completely with one another. For instance, the whole topic of the formation by DNA genes of corresponding messenger RNAs, of the influence of these RNAs on microsomal particles, and of the synthesis by these particles of specific protein molecules, belongs at least as much, if not more, to embryology as it does to genetics. Disputes as to the exact lines of demarcation to be drawn between the subjects are bound to be profitless.

I have brought up this question of the relations between the conceptual fields of genetics and embryology only in order to get it out of the way. The point I want to make is not connected with any alleged difference—which I believe to be unreal—between the theoretical outlooks of embryologists and geneticists. It is not mainly a question of theoretical schemes, but rather of the subject matter to which they are applied; though the nature of this subject matter does, I think, reflect back on the theory in the sense of influencing what range of concepts we consider to be adequate. What I want to urge is that the conventional subject matter of embryology, the developing cells of higher organisms, brings to our attention a whole range of phenomena which fall outside the range normally dealt with by microbiological genetics but which there is no reason to believe can safely be neglected, even when we are searching only for the main outlines of the story.

Consider, for instance, the simplest basic genetic determination, that is to say, the specification by a sequence of DNA nucleotides in the chromosome of a certain sequence of amino acids to be assembled into a protein in a microsome. Now a combination of genetic and biochemical methods applied to microorganisms may carry us a very long way in understanding this matter. The study of genetic recombination, biochemical analyses of amino acid sequences, and perhaps of nucleotide sequences, combined with the use of synthetic nucleotide sequences of known composition, may well, within quite a few years, "crack the genetic code," that is to say, reveal what group of nucleotides in the DNA corresponds to what amino acid in the final protein. This would be—in fact one is almost confident enough to say this will be—a very great triumph indeed; but still it will not be by any means the whole story. We shall want to know, in the first place, how, when, and why a given DNA cistron comes to make a messenger RNA, and secondly how this RNA gets to the microsome. It is in conection with these questions that the cells of higher organisms exhibit phenomena which may yet turn out to be essential components of the whole process of genetic determination.

One of these phenomena is the almost universal occurrence in higher organisms of feedback relations between cytoplasm and genes, such that the nature of the cytoplasm determines the intensity of the syntheses controlled by the various genes in the nucleus. This raises the whole problem of the nature of genotropic substances, which we discussed in the first chapter. They can, of course, also be detected in microorganisms, for instance, in the phenomena of induced enzyme synthesis; but evidence for them is much more widespread in metazoan cells, and it may be suspected that these will prove more favorable material for the biochemical studies which will be necessary to determine their nature.

Another, and presumably allied, phenomenon which is exhibited more definitely in metazoan cells than in microorganisms is the alteration in the reactivity of the genetic controlling system which has been referred to as competence. Microorganisms in which syntheses can be induced seem, so long as they have the basic genetic potentialities, to be always competent to express them as soon as the required inducing stimulus is provided. It is, therefore, difficult to use this material to investigate the conditions on which the reactivity depends. If, for instance, it should turn out that a gene is only reactive to its corresponding genotropic substance when the DNA is combined with a particular type of protein, this could hardly be ascertained except in a system in which the DNA is sometimes combined with something else which renders it unreactive.

Again, we could not claim to have anything like a full understanding of the genetic determination of development until we can describe the way in which the quantities and types of proteins synthesized in metazoan cells are coordinated with one another so that the cell acquires a histologically definite character. This integration of the syntheses may perhaps operate by purely chemical processes of competition, interaction, and so on, between the various synthesizing systems. If this were the case, it could be investigated fully in microorganisms, for instance, by discovering what effect the induction of a particular enzyme has on the other synthetic processes occurring simultaneously in the cell. On the other hand, at present, the actual evidence for such integration is very much weaker in microorganisms than in metazoan cells, and it seems unsafe to neglect the possibility that the integration is related to some of the other obvious characteristics of such cells. I am referring in particular to their high degree of internal structural organization into entities such as the nuclear envelope, ergastoplasm, and the like. We discussed in Chapter 2 some examples of specific ultrastructural organization in different types of cells. This organization involves the entities which we believe to be parts of the genetic effector system, in particular, the nucleus and the microsomes. It is, to my way of thinking, most unplausible to suggest that any widespread characteristics of this effector system will turn out to be not of significance. It seems to be much more likely, though this is admittedly a guess, that the structural organization will be found to be an essential part of the mechanism by which genes become effective.

These considerations bring to our attention the whole problem of morphogenesis. How are we to account for the appearance of structures like mitochondria, chloroplasts, rhabdomeres, etc.? Clearly this is a problem to which strictly genetic methods, such as recombination analysis, etc., have little relevance. Do they fall perhaps within the sphere of biochemistry? Of course, insofar as all structures are made of substances and all substances are chemical entities, morphogenesis of a cellular organelle can be considered as a problem of molecular organization. But I think one must ask the apparently silly-sounding question, "When is a molecule not a molecule?" Platt (1961) has recently written an article with the rather illuminating title "Properties of Large Molecules That Go Beyond the Properties of Their Chemical Subgroups." The forces involved in the organization of lipids, proteins, carbohydrates, etc., into an entity such as the nuclear envelope are quite likely not of the kind usually dealt with by conventional chemistry, even the conventional chemistry of macromolecules. We may be involved here with what might be called supramacromolecular chemistry. This must certainly be true when we go on to consider organization at a slightly greater level of size, such as that involved in the integration of the whole cell into an orderly pattern with its different organelles, nucleus, pigment granules, rhabdomere, etc., arranged in fairly definite geometrical relations with one another. One can hardly deny that in such phenomena one has passed into a realm which is most appropriately considered as an aspect of embryology.

There are then a whole range of problems, such as genotropic substances, competence, cellular ultrastructure, and morphogenesis in general, which do not seem very amenable to the methods either of genetics or biochemistry, at least when those terms are interpreted in a relatively strict sense. It is, as I suggested earlier, not of any great importance whether we decide to widen the meanings of the words "genetics" and "molecule" to cover them, or whether we simply call them "embryology"; but they are, I suggest, problems which it is unsafe merely to leave on one side; and the most favorable material for studying them would appear to be the cells of higher organisms. It is in this sense that I should like to see the present fashion for molecular genetics diluted by the diversion of rather more attention to fundamental embryology. Genetics has had its breakthrough, and those who want quick results can probably get them most easily by exploiting this. But the next breakthrough we need, to round off our understanding of fundamental biological processes, is an embryological breakthrough. Let us hope that we get it soon.

Works Cited

Abbott, U., and M. Kieny. 1961. Sur la croissance in vitro du tibio-tarse et du péroné de l'embryon de Poulet "diplopode." C.R.Acad.Sci.(Paris), 252: 1863–65.

Abercrombie, M. 1961. The bases of locomotory behavior of fibroblasts. Exptl.Cell Res., (Suppl.) 8:188–98.

Afzelius, B. 1959. Electron microscopy of the sperm tail. Results obtained with a new fixative. J.Biophys.Biochem.Cytol., 5:269–78.

Alfert, M. 1958. Variations in cytochemical properties of cell nuclei. Exptl.-Cell Res., 6:227–35.

Anfinson, C. B. 1959. The Molecular Basis of Evolution. New York, Wiley; London, Chapman & Hall.

Anikin, A. W. 1929. Das morphogene Feld der Knorpelbitdung. Arch.f. Entwmech.Org., 114:549–78.

Baglioni, C., and V. M. Ingram. 1961. Four adult hemoglobin types in one person. Nature, London, 189:465–67.

Bairati, A., and F. E. Lehmann. 1952. Über die submicroscopische Struktur der Kernmembran bei *Amoeba proteus*. Experientia, 8:60–61.

Balinsky, B. I. 1959. An electron microscope investigation of the mechanisms of adhesion of the cells in a sea urchin blastula and gastrula. Exptl.Cell-Res., 16:429–33.

Baltzer, F. 1952. Experimentelle Beiträge zur Frage der Homologie. Experientia, 8:285–97.

——— 1957. Über Xenoplastik, Homologie und verwandte stammesgeschichtliche Probleme. Naturforsch.Ges.Bern., 15:1–23.

Barer, R., S. Joseph, and G. Meek. 1960. The origin and fate of the nuclear membrane in meiosis. Proc.Roy.Soc.(London)B., 152:353–66.

Barker, D. C., and K. Deutsch. 1958. The chromatoid body of *Entamoeba invadens*. Exptl.Cell Res., 15:604–10.

Bateman, N. 1954. Bone growth: a study of the gray lethal and microphthalmic mutants of the mouse. J.Anat., 88:212–64.

Beale, G. H. 1954. The Genetics of *Paramecium aurelia*. Cambridge, Cambs.-Univ.Press.

Becker, H. J. 1959. Die Puffs der Speicheldrusen-chromosomen von *Dro-*

sophila melanogaster. I: Beobachtungen zum Verhalten des Puffmusters in Normalstamm und bei zwei Mutanten, *giant* und *lethal-giant larvae.* Chromosoma, 10:654–78.

Beerman, W. 1956. Nuclear differentiation and functional morphology of chromosomes. Cold Spring Harb.Symp.Quant.Biol., 21:217–32.

—— 1959a. Chromosomal differentiation in insects, in Rudnick, ed., Developmental Cytology, pp. 83–103. New York, Ronald Press.

—— 1959b. in Waddington, ed., Biological Organization: Cellular and Subcellular. New York, Pergamon Press.

Benzer, S. 1959. On the topology of the genetic fine structure. Proc.Natl.-Acad.Sci.U.S., 45:1607–20.

Berg, W. E., and W. J. Humphreys. 1960. Electron microscopy of four-cell stages of the Ascidians Ciona and Styela. Devel.Biol., 2:42–60.

Bernal, J. D. 1958. Structural arrangements of macromolecules. Discussions Faraday Soc., 25:7–18.

Bloch, D. P., and H. Y. Hew. 1960a. Schedule of spermatogenesis in the Pulmonate snail, *Helix aspersa,* with special reference to histone transition. J.Biophys.Biochem.Cytol., 7:515–32.

—— 1960b. Changes in nuclear histones during fertilization, and early embryonic development in the Pulmonate snail, *Helix aspersa.* J.Biophys.-Biochem.Cytol., 8:69–81.

Boss, J. M. N. 1959. The contribution of the chromosomes to the telophase nucleus in cultures of fibroblasts of the adult crested newt *Triturus cristatus carnifex.* Exptl.Cell Res., 18:197–216.

—— 1960. The origin of the nucleus after mitotic cell division, in Walker, ed., New Approaches in Cell Biology, pp. 59–66. London and New York, Academic Press.

Brenner, S. 1959. in Waddington, ed., Biological Organization: Cellular and Subcellular. New York, Pergamon Press.

——, F. Jacob, and M. Meselson. 1961. An unstable intermediate carrying information from genes to ribosomes for protein synthesis. Nature, London, 190:576–85.

Brink, R. A. 1958. Mutable loci and development of the organism. J.Cellular-Comp.Physiol., 52:169–96.

Brown, M. G., V. Hamburger, and F. O. Schmitt. 1941. Density studies on amphibian embryos with special reference to the mechanisms of organizer action. J.Exptl.Zool., 88:353–72.

Burnett, F. M. 1959. The Clonal Selection Theory of Acquired Immunity. Cambridge, Cambs.Univ.Press.

——, and F. Fenner. 1949. The Production of Antibodies. London, Macmillan.

Callan, H. G., and L. Lloyd. 1960a. Lampbrush chromosomes, in Walker, ed., New Approaches in Cell Biology, pp. 23–46. London and New York, Academic Press.

—— 1960b. Lampbrush chromosomes of crested newts. Phil.Trans.Roy.Soc.-London B., 243:135–219.

——, and S. G. Tomlin. 1950. Experimental studies on amphibian oöcyte nuclei. I: Investigation of the structure of the nuclear membrane by means of the electron microscope. Proc.Roy.Soc.(London)B., 137:367–78.

Catcheside, D. G. 1960. Complementation among histidine mutants of *Neurospora crassa*. Proc.Roy.Soc.(London)B., 153:179–94.

Chuang, H. H. 1939. Induktionsleistung von frischen und gekochten Organteilen u.s.u. Arch.f.Entwmech.Orig., 139:556–638.

—— 1940. Weitere Versuche über die Veränderung der Induktionsleistungen von gekochten Organtilen. Arch.f.Entwmech.Orig., 140:25–38.

Clayton, R. M. 1960. Labeled antibodies in the study of differentiation, in Walker, ed., New Approaches in Cell Biology, pp. 67–88. London and New York, Academic Press.

——, and T. S. Okada. 1961. Combined influence of antisera and ribonucleic acid preparations on tissue differentiation. Unpublished lecture delivered at Colloque Internationale d'Embryologie, College de France, Paris.

Clever, U., and P. Karlson. 1960. Induktion von Puff-veränderungen in den Speicheldrüsen-Chromosomen von *Chironomus tentans* durch Ecdyson. Exptl.Cell Res., 20:623–26.

Counce, S. J. 1956. Studies on female-sterility genes in *Drosophila melanogaster*. Z.Abstamm.und Vererb., 87:443–92.

—— 1961. The analysis of insect embryogenesis. Ann.Rev.Entomol., 6:295–312.

Crane, H. R. 1950. Principles and problems of biological growth. Sci.Mon., Wash., 70:376–89.

Crick, F. H. C., and A. F. W. Hughes. 1950. The physical properties of cytoplasm. Exptl.Cell Res., 1:37–80.

Crile, G., H. Telkes, and A. F. Rowland. 1932. Autosynthetic cells. Protoplasma, 15:337–60.

Curtis, A. S. G. 1957. The role of calcium in cell aggreration of Xenopus embryos. Proc.Roy.Phys.Soc.(Edinburgh), 26:25–32.

—— 1958. A ribonucleoprotein from amphibian gastrulae. Nature, London, 181:185.

—— 1960a. Cell contacts: some physical considerations. Am.Nat., 94:37–56.

—— 1960b. Cortical grafting in *Xenopus laevis*. J.Embryol.Exptl.Morph., 8:163–73.

—— 1961. Timing mechanisms in the specific adhesion of cells. Exptl.Cell-Res., (Suppl.) 8:107–22.

—— 1962. Cell contact and adhesion. Biol.Rev., 37:(in press).

Danielli, J. F., ed. 1958. Surface chemistry and cell membranes, in Surface Phenomena in Chemistry and Biology, pp. 246–65. New York, Pergamon Press.

Darlington, C. D. 1939. The Evolution of Genetic Systems. Cambridge, Cambs.Univ.Press; Edinburgh, Oliver & Boyd, 2d ed., 1958.

Demerec, M. 1936. Frequency of cell-lethals among lethals obtained at random in the X chromosome of *Drosophila melanogaster*. Proc.Natl.-Acad.Sci.U.S., 22:350–54.

——, et al. 1956. Genetic studies with bacteria. Carnegie Inst. Washington, Publ., 612:1–136.

Dervichian, D. G. 1958. The existence and significance of molecular associations in monolayers, in Danielli, ed., Surface Phenomena in Chemistry and Biology. New York, Pergamon Press.

Dun, R. B., and A. S. Fraser. 1959. Selection for an invariant character, vibrissa number, in the house mouse. Australian J.Biol.Sci., 12:506–23.

Ebert, J. D. 1959. The formation of muscle and muscle-like elements in chorioallantoic membrane following inoculation of a mixture of cardiac microsomes and Rous sarcoma virus. J.Exptl.Zool., 142:587–622.

——, and F. H. Wilt. Animal viruses and embryos. Q.Rev.Biol. 35:261–312.

Edds, M. V., ed. 1961. Macromolecular Complexes. New York, Ronald Press.

Edwards, R. G., and J. L. Sirlin. 1956. Studies on gametogenesis, fertilization and early development in the mouse using radioactive tracers, pp. 376–86. Naples, Proc. 2d World Congr.Fertility and Sterility.

Fabergé, A. C. 1942. Homologous chromosome pairing: the physical problem. J.Genet., 43:121–44.

Fawcett, D. W. 1959. Changes in the fine structure of the cytoplasmic organelles during differentiation, in Rudnick, ed., Developmental Cytology, pp. 161–89. New York, Ronald Press.

Fernandez-Moran, H. 1958. The fine structure of the light receptors in the compound eyes of insects. Exptl.Cell Res., (Suppl.) 5:586–644.

Fincham, J. B. 1959. On the nature of the glutamic dehydrogenase produced by inter-allele complementation at the *am* locus of *Neurospora crassa*. J.Gen.Microbiol., 21:600–11.

Fraser, A. S. 1952. Growth of wool fibers in the sheep. Australian J.Agric.-Res., 3:419–34.

——, and B. M. Kindred. 1960. Selection for an invariant character, vibrissa number, in the house mouse. II. Limits to variability. Australian J.Biol.Sci., 13:48–58.

——, and B. F. Short. 1960. The biology of the fleece. An.Res.Lab.Tech., Paper No. 3., C.S.I.R.O. Melbourne, Australia.

Frey-Wyssling, A. 1948. Submicroscopic Morphology of Protoplasm and Its Derivatives. Amsterdam, Elsevied.

Gall, J. G. 1958. Chromosomal differentiation, in Glass, ed., The Chemical Basis of Development, pp. 103–35. Baltimore, Johns Hopkins Press.

Gay, H. 1956a. Chromosome-nuclear membrane-cytoplasmic interrelations in Drosophila. J.Biophys.Biochem.Cytol., (Suppl.) 2:407–14.

—— 1956b. Nucleocytoplasmic relations in Drosophila. C.S.H.Symp.Quant.-Biol., 21:257–70.

—— 1959. in Waddington, ed., Biological Organization: Cellular and Subcellular. New York, Pergamon Press.

Gibbons, I. R. 1961. Structural asymmetry in cilia and flagella. Nature, London, 190:1128–29.

——, and A. V. Grimstone. 1960. On flagellar structure in certain flagellates. J.Biophys.Biochem.Cytol., 7:697–716.

Glimcher, M. J. 1959. Molecular biology of mineralized tissues with particular reference to bone. Revs.Modern Phys., 31:359–93.

——, 1960. Specificity of the molecular structure of organic matrices in mineralization. Calcification in Biological Systems, pp. 421–87. Washington, Amer.Asso.Adv.Si.

Goldacre, R. J. 1954. Crystalline bacterial arrays and specific long-range forces. Nature, London, 174:732–34.

Goldschmidt, R. B., A. Hannah, and L. K. Piternick. 1951. The podoptera effect in *Drosophila melanogaster*. Univ.Calif.Publ.Zool., 55:67–294.

Goldstein, L. 1958. Localization of nucleus-specific protein as shown by transplantation experiments in *Amoeba proteus*. Exptl.Cell Res., 15:635–37.

——, and W. Plaut. 1955. Direct evidence for nuclear synthesis of cytoplasmic ribose nucleic acid. Proc.Natl.Acad.Sci.U.S., 41:874–80.

Goodwin, B. C. 1961. Studies in the general theory of development and evolution. Ph.D. thesis, Edinburgh University (in press, Academic Press).

Gray, E. G. 1960. The fine structure of the insect ear. Phil.Trans.Roy.Soc.-London B., 243:75–94.

Grimstone, A. V. 1961. Fine structure in Protozoa. Biol.Rev., 36:97–150.

——, R. W. Horne, C. F. A. Pantin, and E. A. Robson. 1958. The fine structure of the mesenteries of the sea-anemone *Metridium senile*. Quart.J.-Micros.Soc., 99:523–40.

Grobstein, G. 1955. Tissue interaction in the morphogenesis of mouse embryonic rudiments in vitro, in Rudnick, ed., Aspects of Synthesis and Order in Growth, pp. 233–56. Princeton, Princeton Univ. Press.

Hadorn, E. 1955. Letalfaktoren. Stuttgart, Thieme. Developmental Genetics and Lethal Factors. London, Methuen, Eng. ed., 1961.

Hamburger, V., and M. Waugh. 1940. The primary development of the skeleton in nerveless and poorly innervated limb transplants of chick embryos. Physiol.Zool., 13:367–80.

Hampé, A. 1959. Contribution á l'étude du développment et de la régulation des déficiences et des excédents dans la patte de l'embryon de poulet. Arch.-Anat.Micr. et Morph.Exptl., 48:347–478.

Hardy, M. H., and A. G. Lyne. 1956. The pre-natal development of wool follicles in merino sheep. Australian J.Biol.Sci., 9:423–41.

Harrison, R. G., K. M. Rudall, and W. T. Astbury. 1940. An attempt at an X-ray analysis of embryonic processes. J.Exptl.Zool. 85:339–56.

Henke, K. 1933. Zur Morphologie und Entwicklungsphysiologie der Tierzeichnungen. Naturwissenschaften, 21:633–90.

—— 1947. Einfache Grundvorgänge in der tierischen Entwicklung. I: Über, Zellteilung, Wachstum und Formbildung in der Organentwicklung der Insekten. Naturwissenschaften, 34:147–57, 180–86.

—— 1948. Einfache Grundvorgänge in der tierischen Entwicklung. II: Über die Entstehung von Differenzierungsmustern. Naturwissenschaften, 35: 176–81, 203–11, 239–46.

Henzen, W. 1957. Transplantationen zur entwicklungsphysiologischen Analyse der carvalen Mundorgane bei Bombinator und Triton. Arch.f.-Entwmech.Orig., 149:387–442.

Hodge, A. J. 1959a. Fine structure of lamellar systems as illustrated by chloroplasts. Revs.Modern Phys., 31:331–41.

—— 1959b. Fibrous proteins of muscle. Revs.Modern Phys., 31:409–25.

—— 1960. Ordnungsprinzipien in der Biologie. Principles of ordering in fibrous systems. Verhandl.Vierter Int.Kong.Elek.Mikr., 2:119–39.

—— 1961. The ultrastructure of the muscle fiber, in Neuromuscular Disorder. Ass.Res.Nerv. and Mental Diseases, pp. 31–51.

——, and F. O. Schmitt. 1960. The charge profile of the tropocollagen macromolecule and the packing arrangement in native-type collagen fibrils. Proc.Nat.Acad.Sci.U.S., 46:186–97.

Holtfreter, J. 1945. Neuralization and epidermalization of gastrula ectoderm. J.Exptl.Zool., 98:161–207.

—— 1947. Observations on the migration, aggregation and phagocytosis of embryonic cells. J.Morph., 80:25.

—— 1948. Significance of the cell membrane in embryonic processes. Ann.-N.Y.Acad.Sci., 49:709–60.

Huxley, H. E. 1961. Muscle cells, in Brachet and Mirsky, eds., The Cell, Vol. IV. New York and London, Academic Press.

Ingram, V. M. 1959. Chemistry of abnormal human hemoglobins. Brit.Med.-Bull., 15:27–32.

Jacob, F., and J. Monod. 1961. Genetic regulatory mechanisms in the synthesis of proteins. J.Mol.Biol., 3:318–56.

Joly, M. 1956. Non-Newtonian surface viscosity. J.Coll.Sci., 11:519–31.

Kallio, P. 1959. The relationship between nuclear quantity and cytoplasmic units in Micrasterias. Ann.Acad.Sci.Fenn., 5–44.

—— 1960. Morphogenetics of *Micrasterias americana* in clone culture. Nature, London, 187:164–66.

Karasaki, S. 1959a. Electron microscopic studies on cytoplasmic structures of ectoderm cells of the Triturus embryo during early phase of differentiation. Embryologia, 4:247–72.

—— 1959b. Changes in fine structure of the nucleus during early development of the ectoderm cells of the Triturus embryo. Embryologia, 4:273–82.

Kendrew, J. C., *et al.* 1960. Structure of myoglobin. Nature, London, 180:422–27.

Kerner, E. H. 1957. A statistical mechanics of interacting biological species. Bull.Math.Biophys., 19:121–46.

—— 1959. Further considerations on the statistical mechanics of biological associations. Bull.Math.Biophys., 21:217–55.

King, R. C., *et al.* 1957. Oögenesis in adult *Drosophila melanogaster*. IV: Hereditary ovarian tumors. Growth, 21:239–61.

——, and R. G. Burnett. 1957. Oögenesis in adult *Drosophila melanogaster*. V: Mutations which affect nurse cell nuclei. Growth, 21:263–80.

—— 1960. Oögenesis in adult *Drosophila melanogaster*. IX: Studies on the cytochemistry and ultrastructure of developing oöcytes. Growth, 24:265–323.

Klein, E. 1960. On the substrate-induced enzyme formation in animal cells cultured in vitro. Exptl.Cell Res., 21:421–29.

Knox, W. E. 1954. In Racher, ed., Cellular Metabolism and Infection, p. 45. New York and London, Academic Press.

Kroeger, H. 1960a. Die Entstehung von Form in morphogenetischen Feld. Naturwissenschaften, 47:148–53.

—— 1960b. The induction of new puffing patterns by transplantation of salivary gland nuclei into egg cytoplasm of Drosophila. Chromosoma, 11:129–45.

Kühn, A. 1955. Entwicklungsphysiologie. Berlin, Springer.

Lamfrom, H. 1961. Factors determining the specificity of hemoglobin synthesis in a cell-free system. J.Molec.Biol., 3:241–52.

Lasansky, A., and E. de Robertis. 1961. Submicroscopic analysis of the genetic distrophy of visual cells in C3H mice. J.Biophys.Biochem.Cytol., 7:679–83.

Leblond, C. P., H. Puchtler, and Y. Clermont. 1960. Structures corresponding to terminal bars and terminal web in many types of cells. Nature, London, 186:784–88.

Le Calvez, J. 1938. Recherches sur les Foraminiferes. I: Devéloppment et reproduction. Arch.Zool.Exptl.Gen., 80:163–333.

Lees, A. D. and C. H. Waddington. 1942. The development of the bristles in normal and some mutant types of *Drosophila melanogaster*. Proc.Roy.-Soc.(London)B., 131:87–110.

Lehmann, F. E. 1945. Einführung in die physiologische Embryologie. Basel, Birkhaüser.

—— 1948. Realizationsstufen in der Organogenese als entwicklungsphysiologisches und genetisches Problem. Arch.Jul.Klaus Stift., 23:568.

Leslie, I. 1961. Biochemistry of heredity: a general hypothesis. Nature, London, 189:260–68.

Lewis, D. 1960. Genetic control of specificity and activity of the S antigen in plants. Proc.Roy.Soc.(London)B., 151:468–77.

Lewis, E. B. 1955. Some aspects of position pseudoallelism. Am.Nat., 89:73–89.

Locke, M. 1959. The cuticular pattern in an insect, *Rhodnius prolixus* Stal. J.Exptl.Biol., 36:459–77.

Lucey, E. C. A., and A. S. G. Curtis. 1959. A time-lapse study of cell re-aggregation. Med.& Biol.Illus., 9:86–92.

McClintock, B. 1956a. Controlling elements and the gene. C.S.H.Symp.-Quant.Biol., 21:197–216.

—— 1956b. Intranuclear systems controlling gene-action and mutation. Brookhaven Symp.Biol., 8:58–74.

Matalon, R., and J. H. Schulman. 1949. Formation of lipoprotein mono-layers: I: Mechanism of absorption, solution and penetration. Discussions Faraday Soc., 6:27–39.

Mauritzen, C. M., and E. Stedman. 1960. Cell specificity of β-histones from the Ox. Proc.Roy.Soc.(London)B., 153:80–89.

Mechelke, F. 1961. Das Wandern des Aktivitatsmaximums im BR_4-locus von *Acricotopus lucidus* als Modell fur die Wirkungsweise eines kom-plexen Locus. Naturwissenschaften, 48:29.

Medawar, P. B. 1950. Transformation of shape. Proc.Roy.Soc.(London)B., 137:474–79.

Mookerjee, S., E. M. Deuchar, and C. H. Waddington. 1953. The morpho-genesis of the notochord in amphibia. J.Embryol.Exptl.Morph., 1:399–409.

Morgan, T. H. 1934. Embryology and Genetics. New York, Columbia Univ.Press.

Moscona, A. A. 1960. Patterns and mechanisms of tissue reconstruction from dissociated cells, in Rudnick, ed., Developing Cell Systems and Their Control, pp. 45–70. 18th Growth Symp. New York, Ronald Press.

Moses, M. J. 1960. Breakdown and reformation of the nuclear envelope at cell division, pp. 230–33. Proc.4th Int.Conf.Electron Microsc., Berlin, Springer.

Nageotte, J. 1936. Morphologie des Gels Lipoides. Paris, Hermann.

Needham, J. 1936. Order and Life. New Haven, Yale Univ.Press.

—— 1942. Biochemistry and Morphogenesis. Cambridge, Camb.Univ.Press.

Neumann, D. 1959a. Morphologische und experimentelle Untersuchungen über die Variabilität der Farbmuster auf der Schale von *Theodoxus fluviatilis L.* Z.Morph.Ökol., 48:349–411.

—— 1959b. Experimentelle Untersuchungen des Farbmusters der Schale von *Theodoxus fluviatilis L.* Verhandl.deut.Zool.Ges.Munster, pp. 152–56.

Okada, E., and C. H. Waddington. 1959. The submicroscopic structure of the Drosophila egg. J.Embryol.Exptl.Morph., 7:583–97.

Palade, G. E. 1955. A small particulate component of the cytoplasm. J.-Biophys.Biochem.Cytol., 1:59–68.

—— 1956. The endoplasmic reticulum. J.Biophys.Biochem.Cytol., (Suppl.) 2:85–98.

——, and P. Siekevitz. 1956. Liver microsomes. J.Biophys.Biochem.Cytol., (Suppl.) 2:171–98.

Pappas, G. D. 1956. The fine structure of the nuclear envelope of *Amoeba proteus*. J.Biophys.Biochem.Cytol., (Suppl.) 2:431–34.

Pasteels, J. 1957. Centre organisateur et potential morphogénétique chez les bactrachiens. Bull.Soc.Zool.France., 76:231.

Pauling, L. 1953. Compound helical configurations of polypeptide chains. Nature, London, 171:59–61.

Pavan, C. 1958. Morphological and physiological aspects of chromosomal activities, pp. 321–36. Proc.10th.Int.Conf.Genet.

—— 1959. In Waddington, ed., Biological Organization: Cellular and Subcellular. New York, Pergamon Press.

Perutz, M. F., *et al.* 1960. Structure of hemoglobin. Nature, London, 185:-416–22.

Picken, L. 1960. The Organization of Cells and other Organisms. Oxford, Clarendon Press.

Pittendrigh, C. S. 1958. Adaptation, natural selection and behavior, in Roe and Simpson, eds., Behavior and Evolution, pp. 390–416. New Haven, Yale Univ.Press.

Platt, J. R. 1961. Properties of large molecules that go beyond the properties of their chemical sub-groups. J.Theoret.Biol., 1:342–58.

Pollock, M. R. 1958. Enzymatic "de-adaptation"; the stability of an acquired character on withdrawal of the external inducing stimulus. Proc.Roy.Soc. (London)B., 148:340–51.

Pontecorvo, G. 1959. Trends in Genetic Analysis. New York, Columbia Univ.Press.; London, Oxford Univ.Press.

Porter, K. R. 1953. Observations on a submicroscopic basophilic component of the cytoplasm. J.Exptl.Med., 97:727–50.

—— 1957. The submicroscopic structure of protoplasm. Harvey Lectures, 51:175–228.

—— 1960a. Problems in the study of nuclear fine structure, pp. 186–99. Proc.4th.Int.Conf.Electron Micros., Berlin, Springer.

—— 1960b. R. D. Machado. Studies in the endoplasmic reticulum. IV: Its form and distribution during mitosis in cells of the onion root tip. J.-Biophys.Biochem.Cytol., 7:167–86.

Raven, C. P. 1958. Information versus preformation in embryonic development. Arch.Neur.Zool., (Suppl.) 13:185–93.

Rebhun, L. I. 1956. Electron microscopy of basophilic structures of some invertebrate oöcytes, I and II. J.Biophys.Biochem.Cytol., 2:93–104; 159–70.

Rendel, J. M. 1959. Canalization of the scute phenotype of Drosophila, Evolution, 13:425–39.

——, and B. L. Sheldon. 1960. Selection for canalization of the scute phenotype in *Drosophila melanogaster*. Australian J.Biol.Sci., 13:36–47.

Reverberi, G., and V. Mancuso. 1960. The constituents of the egg of *Ciona
intestinalis* (Ascidians) as seen by the electron microscope. Acta Embryol.-
Morphol.Exper., 3:221–35.

Rhoades, M. M. 1941. The genetic control of mutability in maize. C.S.H.-
Symp.Quant.Biol., 9:138–44.

Rouiller, C., and E. Fauré-Fremiet. 1957. L'ultrastructure des trichocystes
fusiformes de *Frontania atra*. Bull.Micr.Appl., 7:135–39.

Rudkin, G. T., and S. L. Corlette. 1957. Disproportionate synthesis of DNA
in a polytene chromosome region. Proc.Natl.Acad.Sci.U.S., 43:964–68.

Sager, R. 1958. The architecture of the chloroplast in relation to its photo-
synthetic activities. Brookhaven Symp.Biol., 11:101–17.

Saunders, J. W., J. M. Cairns, and M. T. Gaseling. 1957. The role of the
apical ridge of ectoderm in the differentiation of the morphological structure and inductive specificity of limb parts in chick. J.Morphol., 101:57–
87.

Schleip, W. 1929. Die Determination der Primitiventwicklung. Leipzig,
Akadem.Verlag.

Schmidt, W. J. 1937. Die Doppelbrechung van Karyoplasma, Zytoplasma und
Metaplasma. Berlin, Protoplasma Monogr. 11 Borntraeger.

Schmitt, F. O. 1959. Interaction properties of elongate protein macromolecules with particular reference to collagen (Tropocollagen). Revs.Modern
Phys., 31:349–58.

——, ed. 1960. Fast fundamental transfer processes in aqueous biomolecular
systems. Symposia Dept.Biol., M.I.T.

Schwartz, V. 1953. Zur Phanogenese der Flugelzeichnung von *Plodia interpunctella*. Z.Indukt.Abstamm.Vererb., 85:51–96.

Selman, G. G., and C. H. Waddington. 1953. The structure of the sperma-
tozoa in dextral and sinistral races of *Limnea peregra*. Quart.J.Micros.Sci.,
94:391–97.

—— 1962. The structural potentialities of lipids with particular reference
to the cell membrane: a review. (in preparation)

Sirlin, J. L. 1962. The nucleolus. Prog.Bioph.Biophys.Chemistry, 12: (in
press)

——, S. K. Brahma, and C. H. Waddington. 1956. Studies on embryonic
induction using radioactive tracers. J.Embryol.Exptl.Morph., 4:248–53.

——, and S. K. Brahma. 1959. Studies on embryonic induction using radioactive tracers. II: The mobilization of protein components during induction of the lens. Devel.Biol., 1:234–46.

——, and T. R. Elsdale. 1959. Rates of labeling of RNA and proteins in
cell components of the amphibian myoblast. Exptl.Cell Res., 18:268–81.

——, K. Kato, and K. W. Jones. 1961. Synthesis of ribonucleic acid in the
nucleolus. Biochem.Biophys.Acta, 48:421–23.

——, and G. R. Knight. 1958. The pattern of protein sulphur after Feulgen hydrolysis in the salivary gland chromosomes of *Drosophila melanogaster*. Chromosoma, 9:119–59.

——, and G. R. Knight. 1960. Chromosomal syntheses of protein. Exptl.Cell Res., 19:210–19.

——, and C. H. Waddington. 1956. Cell sites of protein synthesis in the early chick embryo, as indicated by autoradiographs. Exptl.Cell Res., 11:197–205.

Sjöstrand, F. S. 1956. The ultrastructure of cells as revealed by the electron microscope. Int.Rev.Cytol., 5:455–529.

Smith, D. S., and V. C. Littau. 1960. Cellular specialization in the excretory epithelia of an insect, *Macrosteles fascifrons Stål* (Homoptera). J.Biophys.Biochem.Cytol., 8:103–33.

Smith, J. Maynard. 1960. Continuous, quantized and modal variation. Proc.Roy.Soc.(London)B., 152:397–409.

——, and K. C. Sondhi. 1960. The genetics of a pattern. Genetics, 45:1039–50.

Spemann, H. 1938. Embryonic Development and Induction. New Haven, Yale Univ.Press.

——, and O. Schotte. 1933. Ueber xenoplastische Transplantation als Mittel zur Analyse der embryonalen Induktion. Naturwissenschaften, 20:463–66.

Spiegel, M. 1954. The role of specific antigens in cell adhesion. Biol.Bull., 107:49–155.

Steinberg, M. S. 1958. On the chemical bonds between animal cells. A mechanism for type-specific association. Am.Nat., 93:65–82.

Steinert, M. 1958. Action morphogénetique de l'urée sur le trypanosome. Exptl.Cell Res., 15:431–35.

Stern, C. 1954. Two or three bristles. Am.Sci., 42:213–47.

—— 1956. The genetic control of developmental competence and morphogenetic tissue interactions in genetic mosaics. Arch.f.Entwmech.Orig., 149:1–25.

——, and A. Hannah-Alava. 1957. The sexcombs in males and intersexes of *Drosophila melanogaster*. J.Exptl.Zool., 134:533–56.

Stocker, B. A. D., M. W. McDonough, and R. P. Ambler. 1961. A gene determining presence or absence of ϵ-n-methyl-lysine in Salmonella flagellar protein. Nature, London, 189:556–58.

Stumpf, H. 1959. Die Wirkung von Hitzereizen auf Entwicklungsvorgänge im Puppenflügel von Drosophila. Biol.Zbl., 78:116–42.

Swann, M. M. 1961. Unpublished lecture.

Swift, H. 1956. The fine structure of annulate lamellae. J.Biophys.Biochem. Cytol., (Suppl.) 2:415–18.

Szent-Gyorgi, A. 1961. The supra- and submolecular in biology. J. Theoret.-Biol., 1:75–82.

Tartar, V. 1960a. The Biology of Stentor. London, Pergamon Press.

—— 1960b. Reconstitution of minced *Stentor coerulens.* J.Exptl.Zool., 144: 187–207.

Thompson, D'A. W. 1942. On Growth and Form. Camb.Univ.Press, 2d ed.

Toivonen, S. 1940. Über die Leistungsspezifität der abnormen Induktoren im Implantatversuch bei Triton. Ann.Acad.Sci.Fenn.A., 55:1–145.

Townes, P. L., and J. Holtfreter. 1955. Directed movements and selective adhesion of embryonic amphibian cells. J.Exptl.Zool., 128:53–120.

Tschumi, P. 1953. Ontogenetische Realisationsstufen der Extremitäten bei Xenopus und die Interpretation phylogenetischer Strahlenreduktionen bei Wirbeltieren. Rev.Swisse Zool., 60:496–506.

Tuft, P. H. 1962. The uptake and distribution of water in the *Xenopus laevis* embryo. J.Embryol.Exptl.Morph., (in press).

Turing, A. M. 1952. The chemical basis of morphogenesis. Phil.Trans.Roy. Soc.London B., 237:37–72.

Verwey, E. J. W., and J. T. G. Overbeek. 1948. Theory of Stability of Lyophilic Colloids. Amsterdam, Elsevier.

Vogt, M. 1946. Neuer Beitrag zur Determination der Imaginalscheiben bei Drosophila. Experientia, 2:1–5.

Volkin, E., and L. Astrachan. 1957. RNA metabolism of T-2 infected *Escherichia coli,* in The Chemical Basis of Heredity, pp. 686–94. Baltimore, Johns Hopkins Press.

Volterra, V. 1931. Leçons sur la Théorie Mathématique de la Lutte pour la Vie. Paris, Gautiers Villers.

Waddington, C. H. 1934. The competence of the extra-embryonic ectoderm in the chick. J.Exptl.Biol., 11:211–17.

—— 1940a. Organisers and Genes. Cambridge, Camb.Univ.Press.

—— 1940b. The genetic control of wing development in Drosophila. J.Gen. 41:75–139.

—— 1940c. Genes as evocators in development. Growth, (Suppl.): 37–44.

—— 1940d. Somatic mutations of the straw locus in Drosophila. Nature, London, 146:355.

—— 1941. Translocations of the organizer in the gastrula of Discoglossus. Proc.Zool.Soc.A, 111:189–98.

—— 1942a. Observations on the forces of morphogenesis in the amphibian embryo. J.Exptl.Biol., 19:284–93.

—— 1942b. Some developmental effects of X-rays in Drosophila. J.Exptl.-Biol., 19:101–17.

—— 1942c. Growth and determination in the development of Drosophila. Nature, London, 149: 264–69.

—— 1943. The development of some "leg genes" in Drosophila. J.Genet., 45:29–43.

—— 1953. The interactions of some morphogenetic genes in *Drosophila melanogaster.* J.Genet., 51:243–58.

—— 1954. The cell physiology of early development, in Kitching, ed. Recent Developments in Cell Physiology, pp. 105–20. London, Butterworth.

—— 1955. On a case of quantitative variation on either side of the wild-type. Z.Ind.Abst.u.Vererblhre., 87:208–28.

—— 1956. Principles of Embryology. London, Allen & Unwin; New York, Macmillan.

—— 1957. The Strategy of the Genes. London, Allen & Unwin.

—— 1959. Biological Organization: Cellular and Subcellular. New York, Pergamon Press.

—— 1961a. The Nature of Life. London, Allen & Unwin; New York, Atheneum.

—— 1961b. Epigenesis and preformation in the eggs of insects. Proc.Conf. Spallanzani (in press)

—— 1961c. Architecture and information in cellular differentiation, in Ramsay and Wigglesworth, eds., The Cell and the Organism, pp. 117–26. Cambridge, Cambs.Univ.Press.

——, and R. M. Clayton. 1952. A note on some alleles of aristopodia. J.Genet., 51:123–29.

——, and C. B. Goodhart. 1949. Location of absorbed carcinogens within the amphibian cell. Quart.J.Micros.Soc., 90:209–19.

——, and L. Mulherkar. 1957. The diffusion of substances during embryonic induction in the chick. Proc.Zool.Soc.(Calcutta), Mookerjee Mem.Vol., pp. 141–47.

——, J. Needham, and J. Brachet. 1936. The activation of the evocator. Proc.Roy.Soc.(London)B., 120:173–207.

——, and E. Okada. 1960. Some degenerative phenomena in Drosophila ovaries. J.Embryol.Exptl.Morph., 8:341–48.

——, and M. Perry. 1960. The ultrastructure of the developing eye of Drosophila. Proc.Roy.Soc.(London)B., 153:155–78.

—— 1962. The ultrastructure of the developing amphibian notochord. (in press)

——, and E. Okada. 1961b. A note on some structures in Cirratulus eggs. Exptl.Cell Res., 23:634–37.

—— 1961a. "Membrane knotting" between blastomeres of Limnea. Exptl. Cell Res., 23:631–33.

——, and J. L. Sirlin. 1959. The changing pattern of amino acid incorporation in developing mesoderm cells. Exptl.Cell Res., 17:582–84.

Wagner, G. 1949. Die Bedeutung der Neuraleiste für die Kopfgestaltung der Amphibienlarven. Rev.Swisse Zool., 56:519–620.

Waris, H. 1950. Cytophysiological studies on Micrasterias. II: The cytoplasmic framework and its mutation. Physiol.Plantarum, 3:236–46.

—— 1951. Cytophysiological studies on Micrasterias. III: Factors influencing the development of enucleate cells. Physiol.Plantarum, 4:387–409.

Watson, M. L. 1959. Further observations on the nuclear envelope of the animal cell. J.Biophys.Biochem.Cytol., 6:147–56.

Waugh, D. F. 1957. A mechanism for the formation of fibrils from protein molecules. J.Cellular Comp.Physiol., (Suppl. 1) 49:145–64.

—— 1961. Molecular interactions and structure formation in biological systems, in Edds, ed., Macromolecular Complexes, pp. 3–18. New York, Ronald Press.

Weiss, P. 1939. The Principles of Development. New York, Holt.

—— 1950a. The deplantation of fragments of nervous systems in amphibians. J.Exptl.Zool., 113:397–461.

—— 1950b. Perspectives in the field of morphogenesis. Quart.Rev.Biol., 25:177–98.

—— 1956. The compounding of complex macromolecular and cellular units into tissue fabrics. Proc.Natl.Acad.Sci.U.S., 42:819–30.

—— 1958. Cell contact. Int.Rev.Cytol., 7:391–422.

—— 1959. In Waddington, ed. Biological Organization: Cellular and Subcellular. New York, Pergamon Press.

——, and G. Andres. 1952. Experiments on the fate of embryonic cells (chick) disseminated by the vascular route. J.Exptl.Zool., 121:449–88.

Weisz, P. B. 1951. A general mechanism of differentiation based on morphogenetic studies in ciliates. Am.Nat., 85:293–306.

Wettstein, D. von. 1959. Developmental changes in chloroplasts and their genetic control, in Rudnick, ed., Developmental Cytology, pp. 123–60. New York, Ronald Press.

Wigglesworth, V. B. 1940. Local and general factors in the development of "pattern" in *Rhodnius prolixus* (Hemiptera). J.Exptl.Biol., 17:180–200.

Wilde, C. E. 1960. On the enhancement of cellular differentiation in tissue cultures by control of the microenvironment, p. 37. Proc.10th Int.Congr.-Cell Biol.

—— 1961. The differentiation of vertebrate pigment cells. Adv.Morphogen., 1:267–300.

Williamson, M. H. 1957. An elementary theory of interspecific competition. Nature, London, 180:422–25.

Willier, B. H., P. A. Weiss, and V. Hamburger. 1955. Analysis of Development. Philadelphia, Saunders.

Willmer, E. N. 1956. Factors which influence the acquisition of flagella by the amoeba *Naegleria gruberi*. J.Exptl.Biol., 33:583–603.

—— 1958. Further observations on the metaplasia of an amoeba. J.Embryol. Exptl.Morph., 6:187–214.

Wolff, E. 1958. Le principe de compétition. Bull.Soc.Zool.France, 83:13.

Woodger, J. H. 1937. The Axiomatic Method in Biology. Cambridge, Cambs. Univ.Press.

Yamada, T. 1961. A chemical approach to the problem of the organizer. Adv.Morphogen., 1:1–54.

Index